T0290899

THE OTHER BIG BANG

THE OTHER BIG BANG

The Story of Sex and
Its Human Legacy

ERIC S. HAAG

Columbia University Press

New York

Columbia University Press
Publishers Since 1893
New York   Chichester, West Sussex

Library of Congress Cataloging-in-Publication Data
Names: Haag, Eric S., author.
Title: The other big bang : the story of sex and its human legacy / Eric S. Haag.
Description: New York : Columbia University Press, [2024] |
    Includes bibliographical references and index.
Identifiers: LCCN 2024014787 | ISBN 9780231207140 (hardback) |
    ISBN 9780231556835 (ebook)
Subjects: LCSH: Sex—History. | Sex (Biology)
Classification: LCC HQ12 .H328 2024 | DDC 306.709—dc23/eng/20240612
LC record available at https://lccn.loc.gov/2024014787

Cover design and illustrations: Henry Sene Yee

Contents

Introduction

Saudi Women Can Drive But There's Still a Long Road Ahead

Toronto Star, June 25, 2018

Victoria's Secret Casts First Openly Transgender Woman as a Model

New York Times, August 5, 2019

Pay Women's Soccer Players More Than the Men, NWSL Team Owner Says

Washington Post, August 23, 2019

Japan Sends Male Minister to Lead G7 Meeting on Women's Empowerment

Time, June 26, 2023

T he above headlines touch on diverse facets of modern life, from motor vehicle regulations and the fashion industry to diplomacy and elite sports. You may have noticed that they do have one thing in common: they could not have existed were it not for the fact that we humans exist, for the most part, in distinct female and male forms. Zoologists (like me) can easily map these forms onto the definition of sexes we apply across all animals: males are individuals who make sperm, and females are those who make eggs. Everything else that distinguishes our sexes could be regarded as a species-specific set of bells and whistles that help those gametes find their way. We humans really seem to care about those bells and whistles, however, and that is where the tidy categories of zoology run into the messiness of culture.

Arguments about the expectations that surround each biological sex abound. The debates vary in detail from place to place, but they are some of our most contentious. We question what it should mean to be male or female, straight or gay, cis- or transgender. We debate whether gender norms have an innate connection to our sex or are largely societal constructs. We grapple with sexual harassment and whether the half of us who can become pregnant have a right to abort. In some countries, not everyone agrees that women should be taught to read. Next time you pick up a newspaper or visit a news site, note how many stories are ultimately about how we deal with sexes.

A few years ago, it struck me that, while the arguments over sexes and gender are undeniably important (and often rather heated), most of those engaged in them have little sense of the deep history of the very phenomenon about which they argue. We might therefore step back and ask a simple question: Why do we have sexes (and sex) at all? This book is an attempt to answer this question.

Imagine for a moment what wouldn't exist if we lived in an alternative, one-sex world: Two types of bathrooms. Sexism. Mother-in-law jokes. Title IX. Sexual orientation and gender identity. Sex-related inequality and the proposed Equal Rights Amendment. Most sexual violence. Before you roll your eyes and say, "But that's science fiction!," I need to point out that many animals and most plants live in this one-sex world. This does not mean they are asexual—often individuals are hermaphroditic, serving as both mother and father in a sperm-swapping arrangement with their mate. If that works for so many other organisms, how did we get here? The answer provided here, which I hope is at least somewhat satisfying, is essentially a reconstruction of our history. By "history," I don't mean the blip of time that is recorded human history. I mean the inconceivably long evolutionary history of sex itself.

The past events that shaped the sexuality of our species are non-negotiable, and they have produced a complex legacy we still don't understand fully. It would be tempting to follow this observation by suggesting (as some have) that, yes, this is true, and because of it we are under the control of hidden, ancient sex programs that rule our behavior with an iron will. We can't help but engage in infidelity or mansplaining because it's in our genes![1] I see the bait, but as a scientist and a human, I see an awful lot of evidence against this facile claim. That sex is deeply imbedded in our core biology does not make that biology strictly determinative. This seems reasonable, but it prompts a question: How, then, do we acknowledge what is, while not making bone-headed conclusions about what could or could not be? This is a hard question to answer, and I can't fully answer it, but I hope I can provide a perspective that will help other, smarter people do so. Perhaps you are one of them.

I hope the above has convinced you that knowing more about sex, sexes, and their evolution is worthwhile. Because anything with the word "sex" in the title may draw onlookers looking for something salacious, however, I need to issue a warning: you'll find nothing here about G-spots, fetishes, or sexual positions (sorry). Rather, it is a journey through billions of years of evolution, albeit one that at each fork in the road does turn toward our own species. If you haven't just closed the cover and walked away in disappointment, then we can take this trip together.

There are many who could serve as your guide on this tour. I was finally motivated to take on the role because of two main qualifications. First, for three decades I have conducted research on the evolution of animal sex and reproduction. In the process, my collaborators and I have reconstructed how changes in the DNA of a species is connected to changes in the way they develop and reproduce. Whether working with sea urchins, roundworms, or hermaphroditic fish, the goals have always been to understand how new aspects of reproduction appear in evolution and to uncover some of the genetic tweaks that brought them into existence. This work is focused on minute details you probably don't care to hear about, but over time it developed the comparative approach to understanding the origins of our present situation that is applied more broadly in this book.[2]

The second qualification is that of teacher. Since 2002, I have had the privilege of offering introductory and graduate-level biology courses to hundreds of students at the University of Maryland. Over the years, I noticed that the content related to reproduction, sex, and its variation sparked particular curiosity (probably due in equal parts to my enthusiasm for the topic and to the age of the students). In 2017, the opportunity to teach a small seminar course for students not majoring in biology presented itself. The history of sex and its human impact seemed to be of general interest, and in determining what that course should cover, the themes of this book began to coalesce. Their testing in the first iterations of the class provided priceless feedback. I am thus especially grateful to these students and to all my fellow instructors for their unknowing role in this project.

The book is broken into three parts. The first provides evolutionary bookends for the basic biology topics we will encounter. Chapter 1 tells the origin story of sex and how and why sex first began in single-celled organisms. Chapter 2 summarizes human sexual development. These first two chapters describe events separated by about 1.8 billion years of evolution. The second part fills in this huge temporal gap, recounting how sex was modified over hundreds of millions of years in the lineage that leads to primates, the group of mammals that includes us. The evolution of multicellular animals triggered the origin of male and female attributes. These were separated into male and female sexes while our ancestors were still living in the ocean. This second part ends by examining how sex was transformed as our ancestors left water for good and became more mobile.

The third part of the book focuses on humans and our closest kin, the nonhuman primates. We will see that, in some important ways, human sex and reproduction were already set in the ancestor of the great apes, while in others they differ from that of even our closest relatives. With these final pieces of our species' history in place, the last four chapters address a few situations (of the many possible) in which this historical perspective can help us think about issues in contemporary society.

Much of this book relies on a form of historical logic referred to by my colleague David Baum as "tree thinking." We are not talking about botany here but rather the use of branching diagrams called phylogenetic trees. They allow us to pose and evaluate hypotheses about how changes occurred in evolution, changes that collectively led to our current situation. Figure 0.1 presents a tree that emphasizes the ancestors of us humans and some of the other nonhuman descendants of each of those ancestors. The amount of time that any two present-day organisms have been separated is reflected in the length of the branches that connect each to their last common ancestor. In other words, the more closely related are two species, the shorter the path connecting them. With a bit of practice, interpreting such a diagram gets easy.

The tree in figure 0.1 is a visual table of contents of the entire book because each lettered ancestor represents a step on the road to us. This does not mean that the formation of our species was the goal of it all—as far as we can tell, evolution has no goal. But as Neil Shubin has described so well in his book, *Your Inner Fish*, it does mean that we carry inside ourselves features that evolved long ago to address challenges faced by ancestors that lived very different lives. This historical baggage persisted even as new demands shaped how humans undertake sex and reproduction. Our ancestors thus form a chain of foresight-free innovators who, in reacting to their daily struggles, unwittingly set changes in motion that eventually enabled our species (1) to become a dominant organism on the planet (at least from the perspective of habitat usage) and (2) to be easily distracted by two-dimensional images of sexy swimsuit models.

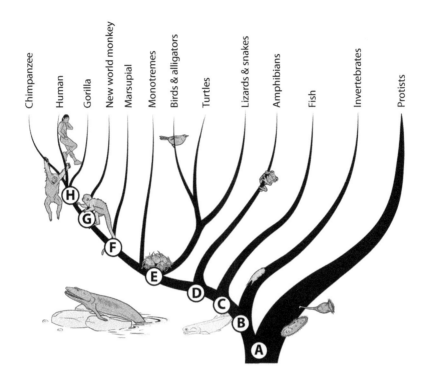

0.1 The book's table of contents as an evolutionary tree. Time moves from bottom to top, with each branch representing a lineage, and each fork reflecting a split in a lineage. Point A represents the ancestor of eukaryotes, the first fully sexual organism and the subject of chapter 1. The diverse array of single-celled eukaryotes are called protists. One group of protists gave rise to the first animals at point B. Chapter 3 describes two major impacts of multicellularity on how sex is implemented in animals: the germline-soma distinction and the origin of male and female. Chapters 4, 5 and 6, deal with the different ways to package male and female traits into organisms, sex determination in male-female species, and the secondary loss of mating. These apply to many animal groups. Most animals are invertebrates that use hermaphroditic or male-female reproduction. Point C is the ancestor of all vertebrates, which had male and female sexes. Most of the descendants of C remained in water (i.e., they are fish). Point D is the first land-dwelling vertebrate, or tetrapod, which laid eggs in water like modern amphibians. Point E is the ancestor of amniotes whose eggs are protected by shells and membranes. The transition from amphibian to amniote is discussed in chapter 7. Point F marks the origin of live birth in mammals (chapter 8). Points G and H are the ancestors of all primates and of the old-world primates (the tail-less monkeys, apes, and humans), respectively. Some aspects of primate reproduction, many of which we humans share, are discussed in chapters 9 and 10. Point I represents the last common ancestor of our species, which acquired several unique features that are discussed in chapters 2, 11 and 13.

Source: Nicholas Bezio and Louisa Wu.

I cannot claim to possess specific training in many of the things covered here, and this is especially true for the final topics. The courage required to plow ahead anyway stems from two sources. First, progress in science and other areas of human activity is often enabled by outsiders sticking their nose into things that are none of their business. For example, the first demonstration that gene sequences could reveal the relatedness of organisms was not by an evolutionary biologist, but by Linus Pauling, a highly decorated physical chemist, and his bio-chemist colleague, Émile Zuckerkandl. Closer to our subject here, Alfred Kinsey owed his faculty position at Indiana University to his expertise on tiny gall wasps (he collected millions of them). He only turned his attention to his own species late in his career, but the Kinsey reports changed the conversation forever.

Then there is the hope of being useful. After five decades of living my per-sonal roles as a son, brother, friend, husband, father, and citizen, I have seen the joy, angst, and sorrow associated with various manifestations of sex and sexes in our species. A few years ago, it occurred to me that for most people on the planet, this is of far greater importance than the research my laboratory con-ducts. You will find little mention of that research here—this is not a "one man's heroic struggle" kind of science book. The work described here has involved a huge number of people of all sorts, working at various times and places around the globe, and they are the real heroes of the story.

As you dive in, please feel free to skip around. Each chapter is intended to be comprehensible on its own, and you may be drawn to some parts of the story more than others. If reading one piques your curiosity about another topic, so much the better. However you may read, it is my hope you will emerge with an appreciation of two main things. First, most of the features of humanity's sexu-ality have ancient roots that greatly predate our existence as a species. Second, reproduction has always been subject to very strong evolutionary forces that in turn produce innate tendencies in both anatomy and behavior. We can wish with all our might that humans were special, that free will and social norms (or prin-cipled rejection of those norms) can give us complete control over our sexuality. A noble idea, but in counting on it, we will be continually disappointed. A better approach, therefore, is to regard self-knowledge as power and to keep our eyes wide open.

Note on Language

A central idea of this book is that, while gendered and sexed categories of humanity do not perfectly fit everyone, they also have a deep history and real meaning that pervades our daily lives. Trans, nonbinary, intersex, gender-nonconforming, and cisgender experiences of sex and gender all have their own meanings and variations. The wildly diverse, now-global community of English speakers has not yet agreed upon how (and even whether) to hold all these people and experiences at the same time. As a result, common English usage often forces human diversity into binaries that exclude people. For example, while the terms "man" and "father" are usually assumed to apply only to biologically male people, and "woman" and "mother" to females, these assumptions don't apply to everyone. At the same time, use of the most precise and maximally inclusive language in all instances (e.g. use of "male parent" or "biological parent who provided sperm" instead of "father") can make writing seem detached from our day-to-day lives. This book therefore attempts to occupy a middle ground. Familiar gendered terms are often used in their traditional sense, but in some cases their limitations are noted, or alternative, more precise language is substituted.

PART I

The Dots That Need Connecting

CHAPTER 1

Sex in the Air

Synopsis

Two billion years ago, photosynthetic air pollution poisoned the atmosphere with oxygen. This promoted the merger of two long-separated prokaryotic lines. The resulting chimeric organism, the first eukaryote, counts us as descendants. Sex emerges as a fix for a mutational crisis resulting from the union.

The year 1968 is largely remembered for its horrors. Before the year was half over, Robert F. Kennedy and Martin Luther King Jr. would both be dead at the hands of assassins, and the Tet Offensive would cost many thousands of lives in Vietnam. At the same time, a more personal drama was unfolding for one Nevada family. Their new baby was born nearly two months premature, and there was a good chance the little guy wouldn't make it. Like many preemies, his underdeveloped lungs struggled to deliver enough oxygen to his bloodstream. An important advance of the 1950s was the discovery that additional oxygen in the incubator greatly increased survival. Too much oxygen, though, and the child could be blinded. In this case, the doctors got the amount just right, and the boy survived with normal vision. He grew into an impulsive child who talked too much, and then into an awkward teenager with an appalling haircut. Eventually he "matured" into an easily distracted adult with a fondness for worms, music, and tacos. That adult would be me, and it's fair to say that I owe a lot to oxygen.

My debt to oxygen in no way makes me special, of course. We all need it. The ceaseless appetite for oxygen stems from the mitochondria that crowd into each of our cells. What is not so obvious, however, is that oxygen hasn't always been in the air. Even less so is that we can thank its arrival for the emergence of sex. To unpack this requires some explanation. Buckle up.

IN THE BEGINNING

4.5 billion years ago, our young planet was so hot that bodies like ours would have simply vaporized. There was not a lot of living happening back then, and certainly no sex. After another half billion years, the crust had finally cooled to below the boiling point of water, and the oceans began to condense. That is, it rained for the first time. And wow, did it rain. The liquid water eventually amounted to about one-thousandth of the planet's entire volume and produced the mix of land and sea on the surface that would look more familiar and hospitable.

Even with a cooler planet and the oceans in place, there would still have been a big problem for organisms like us: there was no oxygen to breathe. The concentration of carbon dioxide was also very high; 2.5 billion years ago, it was in the range of 4 to 5 percent of the atmosphere.[1] To put the impact of that in perspective, we are now facing global climate change as a consequence of increasing carbon dioxide levels from the 0.03 percent of recent preindustrial times to the present 0.04 percent. Methane, an even more potent greenhouse gas that can hang around in the absence of oxygen, was also present at high levels. With that mix, the ancient Earth was still darned hot. It was in this environment, however, life began.

The first major puzzle living cells solved was how to tap sources of energy around them to run their affairs. Just as it takes some effort to tidy up your room, it takes energy to put together a cell's countless orderly structures and to allow the whole to be responsive to its environment. A second need was to obtain the carbon-based building blocks from which the larger, more complex organic molecules of life can be made. Within these two constraints, however, organisms soon evolved to make their living in a variety of ways.

The first cells had to be able to produce their own organic molecules because nobody else was around to eat. These self-feeders, or *autotrophs*, used light or chemical energy to join carbon dioxide from the atmosphere into sugars. The light-powered version is the familiar process of photosynthesis used by algae and plants. The presence of autotrophs provided the opportunity for other organisms to make a living as predators and scavengers, or *heterotrophs* (for "eating another"). This is us: with every slice of pizza you eat, you are proclaiming your own heterotrophic identify. Some cells mixed aspects of photosynthesis and heterotrophy, using light for energy but relying on other organisms to provide fixed carbon.

These variable chemical preferences influenced where organisms could live. The environments suited for some strategies could be lethal to those using others. As life spread, autotrophs and heterotrophs evolved complex interactions. The result was the Earth's first ecosystems. There is no sign, however, that these included the *eukaryotes*, the organisms who pack their cellular DNA into nuclei. Eukaryotes include animals, plants, fungi, and the thousands of one-celled organisms traditionally called protists. They are the central characters of this book because only eukaryotes have full-blown sexual reproduction. As we will see, it took an unprecedented cataclysm to bring both eukaryotes and their sex lives into existence.

OXYGEN: THE FIRST AIR POLLUTION

Cyanobacteria, named for their blue-green color, are individually microscopic. Yet these tiny microbes played a huge role in the history of life on Earth. We know this in part because some cyanobacteria mix with layers of rock and sand to form large colonial blobs called stromatolites. Stromatolites are the oldest fossils known, dating back at least 3.5 billion years.[2] They are not drawing crowds at the museums like, say, a *T. rex* would, but they provide evidence that cyanobacteria set in motion one of the most dramatic changes our planet ever experienced, and one without which dinosaurs, giant squid, and sex could never have evolved. The next time you enjoy *Spirulina* (another type of cyanobacterium) in your smoothie, you really should thank them.

The evidence for how cyanobacteria changed the world is literally ironclad. Away from oxygen, the Earth's iron is found as it was first deposited when the planet formed, in the water-soluble *ferrous* form. In the presence of oxygen, however, it reacts to form *ferric* iron oxide, or rust. In the early, oxygen-free world, ferrous iron was everywhere, both in surface rocks and dissolved in the seas. In rocks formed from the ocean floor around 3.8 billion years ago, however, geologists see layers of insoluble iron appear for the first time. These red bands tell us that something was producing free oxygen, converting the iron to the ferric form and making it fall out of solution. This rain of rust formed vast deposits of iron-rich minerals, some of which are mined today. The iron bands' worldwide distribution tells us that the process was happening everywhere at once. Photosynthetic organisms could explain these observations, and the near-coincidence in the geological record of banded iron with the first stromatolites clinches it. The world was rusting, and cyanobacteria were responsible.

For something like 1 billion years, the reaction of oxygen with iron acted like a great chemical sponge, removing oxygen as fast as cyanobacteria produced it. Eventually, however, the ferrous iron in the seas and surface rocks was exhausted. Still, the cyanobacteria kept on producing more oxygen. A little over 2 billion years ago, it began to accumulate in the atmosphere for the first time (figure 1.1). From near nothing, oxygen shot up to about 4 percent of the atmosphere in a "mere" 500 million years. This is only one-fifth of today's level, but still a threat to many organisms that had lived for eons without it. The Great Oxygenation Event had arrived, and the world would never be the same.

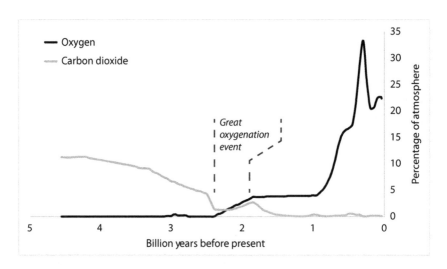

1.1 The Great Oxygenation Event. Initially devoid of oxygen and rich in carbon dioxide, photosynthesis by cyanobacteria flipped the atmosphere's composition to be rich in oxygen.

WHAT YOU LEARNED IN SCHOOL, V. 1.0

As a kid in the 1970s, I learned that there were two fundamental types of organisms on Earth: those whose cells had nuclei, mitochondria, and other membrane-enclosed parts inside them (the eukaryotes) and those that did not, or the bacteria, or *prokaryotes*.[3] Biologists at the time wondered: were eukaryotes present all along, yet somehow absent from the fossil record, or did they actually descend from prokaryotes, with radical modifications along the way? A crucial part of the answer was the gradual realization that the mitochondrion was not just another part of the eukaryotic cell.

In the early 1950s, the memory of the atomic bombs used over Hiroshima and Nagasaki was still fresh, and the long-term consequences for survivors was still unclear. As the Cold War intensified, fear of nuclear war led to both backyard fallout shelters and a scientific push to understand the effects of radiation on living cells. In 1952, Shigefumi Okada came from Japan to the University of Rochester for graduate school on a Fulbright fellowship. Okada knew that cells responded to radiation by producing an enzyme, DNase II, that cuts DNA into pieces. Which part (or parts) of the cell was responsible? His first candidate was the mitochondrion, which could be purified and studied in isolation. He got some from rat livers and irradiated them with X-rays and gamma rays.[4] Okada found that the damaged mitochondria released DNAse II, and in 1957, he reported that they also leak a DNA-like substance. The nucleus was known to be the seat of heredity, and Okada's Rockefeller University colleague, Oswald Avery, had demonstrated that DNA was the hereditary material a few years earlier. Why would mitochondria also contain DNA and an enzyme whose only known function is to cut it? A good question, but Okada was apparently more interested in the damage per se, and didn't pursue it.

The next clue came from across the Atlantic in 1963. The Swedish wife-husband team of Margit and Sylvan Nass were experts in the then-new technique of transmission electron microscopy (TEM). TEM can visualize far tinier structures than possible with any light microscope and relies on chemical stains to highlight different kinds of molecules. The Nasses found that heavy metals that stain DNA marked not only the expected nuclear chromosomes of eukaryotic cells but also (if you look really carefully) small fibers in their mitochondria.[5] They bravely conjectured that "the unavoidable conclusion appears to be that the mitochondrial fibers contain DNA." Further evidence was provided in 1964 by another husband and wife team, Loyola University's Edmund and Sophie Guttes. They fed cells a radioactive precursor of DNA and found that it accumulated both in the nucleus and in mitochondria.[6] As evidence for a genome inside mitochondria mounted, other researchers found that chloroplasts, the photosynthetic organelles of algae and plants, also harbor their own genomes.

The American microbiologist Lynn Margulis made a major contribution by tying these and other observations together. Margulis's University of Wisconsin professor, Hans Ris, had proposed that perhaps the chloroplast used to be a conventional, free-living cyanobacterium. That would mean it was somehow incorporated stably into eukaryotic cells, a process called *endosymbiosis*. With the discovery of mitochondrial DNA, Ris's suspicions of chloroplast endosymbiosis, and her own examination of microscopic cell structures, Margulis made the final intuitive leap in 1967: the simplest explanation unifying everything is that the mitochondrion was once a separate living cell, with its own genome.[7] Although she became famously iconoclastic in her later years, this contribution cemented Margulis's place in history.

In 1981, the entire human mitochondrial genome was sequenced. It is tiny in comparison to that of a free-living bacterium (20,000 base pairs versus approximately 5 billion) and contains only three dozen genes.[8] Nevertheless, its circular form and bacterial-style gene structure made its origins abundantly clear. By comparing the mitochondrial sequences to all known living bacteria, a striking similarity was found with today's alpha-proteobacteria. These heterotrophs consume oxygen and release carbon dioxide, just as eukaryotes (enabled by mitochondria) do today. Although the nature of the cell that internalized the mitochondrion remained inscrutable, it was already clear that the symbiotic organism owed its oxygen-tolerating lifestyle to a bacterial endosymbiont that already possessed it.

We seem to have drifted away from the topic of sex. Why is the mitochondrion's origin relevant? We need one more piece of the story to see why.

WHAT YOU LEARNED IN SCHOOL, V. 2.0

The ability of gene sequences to reveal evolutionary relationships independent of outward appearance was revolutionary for microbiology. University of Illinois microbiologist Carl Woese and his colleagues were quick to see the potential.

In the 1970s and 1980s, they painstakingly sequenced the RNA molecules that form the heart of ribosomes, the protein-synthesizing machines inside cells. These molecules were chosen because they were relatively easy to purify in large amounts and were known to be present in all living things (and thus comparable). Direct sequencing of ribosomal RNA using methods that were then available was arduous and, given today's technology, seems akin to filling bags of sand one grain at a time. Hard though the work was, it was worth it.

Woese discovered that hiding within the traditional bacteria were two distinct types of ribosomal RNA sequences and, by inference, two distinct types of life. In particular, the prokaryotes that produced methane formed their own domain, one as different from other bacteria as the eukaryotes were. He dubbed this new group the Archaea.[9] Although the distinct history of Archaea was new to science, their existence was long known: Archaea in our gut produce the methane in our farts.

Since their discovery forty years ago, new groups of Archaea have been identified. Not all produce methane, but most live in what we humans would regard extreme environments. A key feature of these environments is often a lack of oxygen. While some bacteria also live in low-oxygen (anaerobic) habitats today, the Archaea are generally restricted to them. More genome sequencing revealed that the barrier between the Bacteria and Archaea was somewhat porous: bacterial genes were occasionally acquired by Archaea, allowing them to learn new tricks along the way. But a core set of "information processing" genes, such as those related to the ribosome, remained distinct.

By the 1980s, most biologists were comfortable with the idea that the Bacteria, Archaea, and Eukarya were three descendant lineages from a very ancient split. The standard model for how modern eukaryotes evolved became this: an ancient eukaryote of unknown lifestyle internalized a bacterium that gave rise to the mitochondrion. Around the same time (or perhaps somewhat before) the nucleus and sexual reproduction appeared. Later, all remaining eukaryotes still lacking a mitochondrion went extinct. A subset of eukaryotes went on to a second round of endosymbiosis, acquiring the chloroplast by internalizing a cyanobacterium. These became the algae and (eventually) plants. If you took introduction to biology in the 1990s or later, you likely learned this three-domains-plus-endosymbiosis model (figure 1.2).

In retrospect, there was some eukaryote chauvinism implicit in the three-domain model. The eukaryotic domain of life includes us and every other multicellular

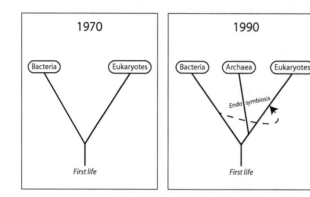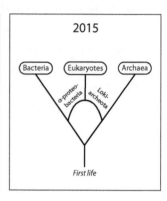

1.2 Changing views of the relationships between prokaryotes and eukaryotes.

organism, that is, all the charismatic macroscopic life that fills the pages of *National Geographic*. I recall having the impression in the 1990s that the hypothetical pre-endosymbiosis eukaryote was driving the evolutionary bus and that the mitochondrion was just a hapless, kidnapped stooge. *Those poor prokaryotes remained simple and never really amounted to much, did they?* Nagging questions remained, however: what exactly was that hypothetical eukaryote before it internalized the mitochondrion, and how did it live? The answer would take another thirty years to sort out.

THREE BECOMES TWO

Within the three-domain model, it became important to resolve the relative timing of the separation of the Bacteria, Archaea, and Eukarya. Did they part ways more or less simultaneously, or might two of them have a more recent common ancestor? These events took place very far in the past, and the signal of relatedness in DNA sequences is gradually drowned out by noise as time passes. One of the few approaches available here is to examine slowly changing genes that had duplicated before the three groups diverged. An example is the gene pair *EF-Tu* and *EF-G*, which encode core parts of the protein-translating ribosome. By using one copy as a common reference standard to measure relatedness in the other, a tentative answer was found: Eukaryotes and Archaea were slightly more closely related to each other than either was to Bacteria.[10]

What? Indeed, the intuitive expectation that the two prokaryotic domains should be more closely related was not holding up. The three-dimensional structures of the ribosome seemed to tell a similar story.[11] Although some other research did support the Bacteria and Archaea as closest relatives, a deep connection between Archaea and Eukarya was emerging. Nevertheless, our specialness still seemed safe: archaea always had other archaea as their closest relatives, and eukaryotes other eukaryotes, so the clean separation between these domains appeared to be solid.

In 2015, however, a new discovery required us to rethink what being a eukaryote means. Swedish microbiologists led by Thijs Ettema and Lionel Guy reported their discovery of new Archaea that live beneath the seafloor of the Arctic Mid-Ocean Ridge, roughly halfway between Greenland and northern Norway. The ocean is over three kilometers deep here and just above freezing, yet life nevertheless abounds. By sequencing DNA from the organisms that live in the cold muck, they found genomes that were clearly from the Archaea but distinct from others previously known. They dubbed them the Lokiarchaeota.[12]

The Lokiarchaeota form part of a larger set of newly discovered Archaea, the Asgard superphylum. When Asgard sequences were added to the family tree of Archaea, something crazy happened: the previously clean separation between Eukarya and Archaea melted away. The analyses indicated that Eukarya are actually a specialized branch emerging from within the variety of Archaea. The closest archaeal relatives to us eukaryotes are the Lokiarchaeota. Consistent with this, Lokiarchaeota contain genes encoding proteins thought previously to exist only in eukaryotes, such as tubulin and actin, which allow our cells to change their shapes dynamically.[13]

As I first drafted this chapter in early 2019, the Asgard organisms only existed as uncultured, disembodied genome sequences. That summer, however, a large team of Japanese biologists led by Hiroyuki Imachi and Masuaru Nobu reported the first lab cultures of Lokiarchaeaota from marine sediments, and later that year, they published their findings.[14] These were heroic experiments, requiring years of careful incubations in anaerobic vessels supplemented with methane. The effort paid off. The team was able to enrich their cultures gradually for the elusive organism, dubbed *Prometheoarchaeum* until they could image it with a microscope.[15] As expected, it is unlike any other prokaryote known. Especially impressive is that *Prometheoarchaeum* can produce elaborate protrusions from its cell surface, the kind that were thought to be restricted to eukaryotes. Such protrusions may have also enabled engulfment of the mitochondrion 2 billion years before.

The discovery of the Lokiarchaeota tells us that the cell that engulfed the mitochondrion was not from a lineage distinct from prokaryotes but rather a bona fide archaeon. The eukaryotic domain of life of which we are part is not a lopsided partnership between a sophisticated eukaryote and a hapless captive but the fusion of the two prokaryotic domains (figure 1.2). Put another way, we smug humans are really just colonial Archaea with endosymbiotic Bacteria inside our cells. The discovery of the Lokiarchaeota has transformed our understanding of the history of life and merits a book of its own. I provide this short account here, however, because the identity of the partners that joined together provide an important clue about why sex appeared in eukaryotes.

The fact that eukaryotes came into existence only with the mitochondrial endosymbiosis allows us to reconstruct the impetus for the merger more confidently. Lokiarchaeota are all anaerobic, and their ancestors of 2.5 billion years ago were living large in the oxygen-free world. Yet cyanobacteria were quietly at work, and the Great Oxygenation Event was gradually relegating Archaea to the remaining habitats that lacked oxygen. Most stayed there, hiding in mud under lakes and oceans, in hot springs (where water retains little dissolved oxygen), and in our own guts. In such a world of restricted habitat options, a ticket out would be a game changer.

As the oxygen accumulated in the atmosphere, some Bacteria found a way to make lemonade out of lemons: they incorporated oxygen into their metabolism. More precisely, they evolved to use oxygen as a new component in their *electron transport chain*. This complex of proteins, imbedded in the cell membrane of all prokaryotes, captures energy from light or food and stores it in the form of adenosine triphosphate (ATP). To achieve this, electron transport chains all require a specific molecule as their end point, the *terminal electron acceptor*. In the beginning, oxygen was not available to serve this function. Once it was, however, aerobic bacteria began to use it. Not only did using oxygen as the terminal electron acceptor turn a poison into a useful tool, it was superior: oxygen allows capture of a great deal more energy from food than other available acceptors.

It is possible that eventually one or more groups of Archaea would have independently evolved the use of oxygen in their metabolism in the same way that the Bacteria did. Instead, like purveyors of counterfeit Gucci handbags and pirated DVDs, a clever Lokiarchaeotan swiped another organism's intellectual property. While it has been suggested that the partnership between Archaea and Bacteria may have begun prior to the Great Oxygenation Event,[16] without a doubt its most important consequence was to allow an archaeon to acquire oxygen tolerance and emerge into the wider world.[17]

Perhaps the protomitochondrion was a partner in a small ecosystem that protected the archaeon from oxygen by consuming it. Or perhaps the protomitochondrion may have been an uninvited guest at first: some modern alpha-proteobacteria live inside other cells as parasites. These include germs that make us humans sick, such as *Rickettsia* (the cause of typhus and Rocky Mountain Spotted Fever). Finally, the aerobic bacterium may have been internalized against its will but was fortuitously prepared to survive. In any case, it persisted. Its archaeon host soon lost its own anaerobic electron transport chain, becoming utterly dependent on the internalized bacterium in the process (figure 1.3).

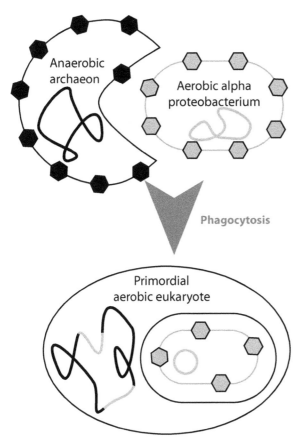

1.3 Endosymbiotic origin of the mitochondrion. An archaeon with an anaerobic electron transport chain (ETC, black hexagons) engulfed an aerobic alpha-proteobacterium that used oxygen as the terminal acceptor in its own ETC (gray hexagons). Both prokaryotes had circular genomes (black and gray squiggles). Internalization by phagocytosis left a host-derived membrane (black) around the endosymbiont (gray). Later, the anaerobic ETC was lost and the mitochondrial genome shrank as DNA was lost or transferred to the host genome. At some point, the host genome became linearized (not shown).

If the mitochondrion used to be a bacterium, why is its genome so small? Much of it was simply lost because many of its genes were no longer relevant inside a host cell. More surprising, however, is that many genes encoding proteins that continued to function in the mitochondrion were transferred to the chromosomes of the host cell.[18] How this occurred is still somewhat mysterious, but it was likely made possible by the many copies of the mitochondrial genome present in each cell. Rupture of a mitochondrion releases its DNA to the cytoplasm, and during cell divisions, this could mingle with chromosomal DNA. Fragments of mtDNA might occasionally be linked to the chromosomes in the process.

THE BIRTH OF VENUS

The first eukaryotes enjoyed lifestyles that would have killed their archaeal ancestors. Their descendants eventually spread throughout the oceans, into fresh water, and onto land. They include all the truly multicellular forms of life (the subject of the next chapter). This freedom came at a price, however: electron transport chains that use oxygen produce toxic by-products. These include hydrogen peroxide and other, more exotic *reactive oxygen species* (ROS). While an aerobic bacterium can release most of its ROS harmlessly at the cell surface, once trapped inside another cell as the mitochondrion, the ROS has nowhere to go. Indeed, nearly all the ROS inside our cells come from mitochondria.

An analogy comes to mind at this point. Imagine your power goes out in a storm, and the only way to get electricity is to use your gas-powered car as a generator. There is hail and torrential rain outside, so you decide to bring the car into the garage and run it in there. Now you can charge your phone, stay warm, and listen to the radio! But wait. We know that as awesome as that seems at first, running a car in a closed garage will trap carbon monoxide. The next thing you know, your cat is dead, and you aren't feeling so well yourself. Just before you lose consciousness, it becomes clear that this great idea came at a steep price that took some time to become apparent.

Internalizing the mitochondrion greatly expanded the available habitat of the host archaeon, but accumulated ROS began to damage the molecules that it needs to live. While ROS damage to proteins, lipids, and RNA molecules are problematic, damage to DNA is especially so. As the long-term repository for the instructions for producing new protein and RNA molecules, DNA cannot easily be replaced. A change in the genome's DNA sequence (a *mutation*) can inactivate a gene. If the mutation cannot be quickly recognized and repaired, the cell could die. While it is true that some mutations are beneficial, far more are harmful. For this reason, any time an organism is hit by a large number of mutations, as occurs with acute radiation exposure, the results are generally lethal. This is so effective that a sealed carton of milk can be completely sterilized by the DNA-damaging effects of gamma rays, remaining drinkable on the shelf at room temperature for years.

The presence of multiple mitochondria deep inside early eukaryotic cells likely generated an unprecedented ROS burden and with it a shotgun blast of new mutations. These would have hit all over the genome, and mechanisms that had long sufficed in the Archaea could no longer reliably rid the genome of harmful variants fast enough to avoid doom. Aerobic bacteria had countered the threat of ROS by evolving enzymes that could detoxify them. They also had invented a limited form of sex, in which small pieces of the genome could be exchanged with other cells. The enzymes that detoxify ROS and that mediate this partial genetic exchange were carried over into eukaryotes, but they were evidently not enough. As detailed below, eukaryotic sex elaborated prokaryotic recombination to the scale of the entire genome. The most plausible explanation for the coincidence of this innovation with the origin of eukaryotes is a DNA damage crisis. Sex provides a way to make a functional genome out of highly damaged starting materials reliably.[19]

What sex achieves would be appreciated by a cash-strapped aficionado of vintage American cars. Perhaps she dreams of driving a 1957 Chevrolet Bel Air, a glorious space-age machine of fins and chrome, but she cannot afford a working car. Undeterred, she buys an inoperable one at a local junkyard. If that car required only one or two part replacements to run, she could buy a few things and be on her way. But suppose the best car she could find had many problems, and some parts were hard to get. By buying a second junker car and carefully swapping parts, she eventually produces a functional vehicle (and a second one that is *really* junk). The damaged genomes of two prospective mates are like the two cars—neither is in good shape. By creating novel combinations of genes, an offspring far healthier than either parent can be produced.

Sex achieves new gene combinations through the alternation of two eukaryote innovations: *syngamy* and *meiosis*. Syngamy, also called fertilization, is the fusion of two *haploid* cells, that is, cells bearing single, distinct versions of the genome. This forms a new cell, called the *zygote*, that now has a pair of genomes.

Cells that have two copies of the genome, like nearly all those in your body, are *diploid*. Only these diploid cells have the option to enter meiosis, which scrambles the two copies again to make unique haploid progeny. This process is summarized in figure 1.4. The first step is to duplicate both copies so that there are

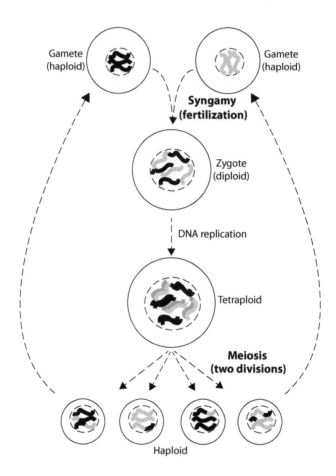

1.4 The cycle of sex. At the top, cells have one copy of their genome, composed of three linear chromosomes. Those able to fuse with other cells are termed gametes. Fusion forms a diploid zygote, with two distinct copies of the genome. Meiosis reverses syngamy but with variation. After each parental chromosome is copied (forming a temporary tetraploid), one copy exchanges a segment with its equivalent from the other parent, while the other two copies remain unchanged. Two meiotic divisions reduce the copy number of each chromosome to one with independent assortment, producing four unique haploid cells unlike either of the starting gametes. The nonsexual division cycles (mitoses) of the haploid or diploid phases are not shown here.

four total. Next, over a series of two consecutive cell divisions, the four copies of the genome are shuffled and dealt out into four haploid cells. Each has one copy of each chromosome but in a unique combination not seen in either of the original two copies.

Two aspects of meiosis contribute to its unique end products. First, the copies of the two starting chromosomes are dealt out without regard to the source of the others, a process referred to as *independent assortment*. Independent assortment is like dealing out one red playing card for each numerical value in the deck of cards but not paying attention to whether it was a heart or a diamond. The variation possible from this mechanism alone is staggering: Each human parent can pro-duce over 8 million unique combinations from their two sets of chromosomes. Mating between two unrelated people samples from all possible pairwise combi-nations of those, an 8-million × 8-million grid that works out to over 70 trillion unique diploid possibilities.

Seventy trillion combinations are impressive, but we're not done yet. Every meiotic cycle, counterpart chromosomes from the two distinct sets swap pieces, or *cross over*, typically at one point along their length. This would be like taking a three of hearts and three of diamonds, cutting them in half, swapping the tops and bottoms, and taping them together to make entirely new cards. Crossing over means the number of distinct haploid products of meiosis is far greater than that from independent assortment alone, and the number of unique diploid combinations becomes effectively infinite.

THE SEXUAL TOOL KIT

Meiosis and syngamy depend on proteins with very specific jobs. Where did they come from? As is generally the case in evolution, they did not fall from the sky in an act of cosmic providence. Rather, they can be traced to antecedents in the two prokaryotic lineages that gave rise to eukaryotes.

The essence of syngamy is the fusion of two haploid cells to become one zygote. Working in lilies and the mustard relative *Arabidopsis*, Japanese and American biologists independently discovered that a gamete surface protein called HAP2 has an essential role in this fusion.[20] Eukaryotes as distantly related as algae, the malaria parasite *Plasmodium*, and jellyfish were subsequently all found to depend on HAP2 for fusion of their gametes.[21] Because it is found in such diverse eukaryotes, HAP2 must have been a component of the primordial sexual tool box. However, some eukaryotes (including vertebrate animals) replaced HAP2 with an alternative fusion-promoting system for reasons that are not at all clear. Evolution is littered with similar cases of one solution being replaced by another, indicating that things that aren't broken are often "fixed" anyway.[22]

Where did HAP2 come from? It appears that we have our historical enemies, the viruses, to thank for that. When the three-dimensional structure of HAP2 was determined in 2017, it bore striking similarity to class II viral fusion proteins.[23] Viruses that have a membrane envelope around them, such as the nasty human pathogens dengue and Zika, rely on these to unlock the door of the cell they are infecting. Imbedded in the viral envelope, the fusion proteins distort the would-be host cell's surface membrane until a porthole opens between the virus and target cell. This quickly expands, dumping the contents of the virus to the interior of the cell. This is fundamentally similar to syngamy, and it is very likely that primordial eukaryotes acquired HAP2 from an ancient virus. In doing so, they co-opted a weapon of their tiny enemies for their own uses.

The origins of meiosis can also be inferred from comparisons with other proteins. The DNA cutting and pasting aspects of meiosis require numerous enzymes, and these have clear counterparts in prokaryotes (table 1.1). Archaea and Bacteria use these to recognize and correct DNA damage. They also occasionally engage in a form of partial sex (termed *conjugation*) in which some DNA is imported via a brief connection with another cell, or even from the environment.

TABLE 1.1 Enzymes involved in meiosis are also found in prokaryotes

Prokaryote Name(s)	Eukaryotic Name	Function
Topoisomerase VIA	*SpoII*	Cuts DNA strands to begin recombination between chromosomes
Rad50	*Rad50*	Trims one DNA strand to promote recombination (works with *MreII*)
MreII	*MreII*	Trims one DNA strand to promote recombination (works with *Rad50*)
SSB	*RPA*	Single-stranded DNA-binding protein
RecA, RadA	*Rad51*	DNA strand exchange between chromosomes
RuvAB	*Hel308*	Resolution of DNA crossovers

Source: M. F. White, "Homologous Recombination in the Archaea: The Means Justify the Ends," *Biochemical Society Transactions* 39, no. 1 (January 2011): 15, https://doi.org/10.1042/BST0390015.

This ability to take up DNA is powerful and enables the spread of resistance to antibiotics, for example. This indicates that prokaryotes did not need eukaryote-style sex for either DNA repair or adapting to new circumstances. What changed to compel early eukaryotes to produce new variants over their entire genomes? The most obvious motivation was protection from DNA damage from the mitochondrion, although there may have been others.

Sex may have originated as a supercharged repair mechanism, but once established, it had other benefits. For example, imagine that a beetle could exploit a plentiful berry as a food source if it were to evolve both red wing covers (to help it blend in) and a key enzyme that enables digestion. If each trait requires a distinct genetic variant to evolve, the only way to get both traits without sex is for an individual to acquire them sequentially. A new red mutant would have to appear, and then its descendants must hang on long enough to experience a mutation that enables the digestive adaptation, or vice versa. This is a very slow waiting game and would often fail. With sex, however, individuals bearing each of the two traits can create an offspring that has both traits in one or two generations. Both theory and lab experiments confirm that this adaptation-speeding benefit of eukaryote sex is real.[24] Evolutionary biologists have long argued about whether sex exists primarily for DNA repair or as a means to facilitate adaptation.[25] In 2024, the answer is clear: it's both.

PUTTING SEX INTO PRACTICE

Even after the basic alternation of meiosis and syngamy was established in eukaryotes around 2 billion years ago, most cell divisions continued to use mitosis. This produces exact genetic copies of the starting cell (regardless of whether there are one or two copies of the genome present). But eukaryotes work sex into their life cycles in various ways. Many spend most of their lives as haploids, dividing by mitosis again and again before merging with other cells to make a diploid zygote. For other organisms (like us), the diploid cells that are formed by fusion divide many times before meiosis. Meiosis (for diploids) or syngamy (for haploids) is often induced by a trigger, such as stress or a change in the season.

Diatoms are diploid algae that have a particularly peculiar cue for inducing meiosis. Each diatom cell is protected by a beautiful skeleton called a *frustule*, a pair of overlapping cups made of silica glass (figure 1.5). Frustules are natural

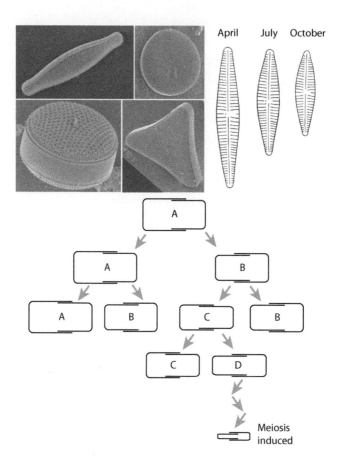

1.5 Growth and the regulation of sex in diatoms. Upper left: scanning electron microscope images of various diatom frustules collected from "marine snow" at a depth of 400 meters. Upper right: progressive shrinkage of descendants of a single sexual auxospore observed by Lothar Geitler in 1932. Bottom: depiction of size reduction process and induction of meiosis. Letters indicate equivalent size cells, with A representing the largest and subsequent letters progressively smaller sizes. Sex can be induced once cells get below a threshold size.

Source: (*Upper left*) Scanning electron micrographs by Timothy Maugel (University of Maryland); (*upper right*) modified from L. Geitler, *Der Formwechsel der pennaten Diatomeen* (Jena: Gustav Fischer, 1932); (*bottom*) redrawn from S. R. Laney, R. Olson, and H. M. Sosik, "Diatoms Favor Their Younger Daughters," *Limnology and Oceanography* 57, no. 5 (2012): 1572–78, https://doi:10.4319/lo.2012.57.5.1572

nanoparticles, so abundant they are mined and used in products as diverse as dynamite and potting soil. Each daughter makes a new half-frustule, but this must fit inside the original. In the mid-nineteenth century, Sir John Denis Macdonald, a surgeon in the British navy with an unexpected penchant for diatoms, first realized that this necessarily reduces the size of one daughter cell every division.[26] Around the same time, the idea also occurred to the German botanist Ernst Pfitzer, who more precisely illustrated the expected consequences.[27]

The progressive shrinkage idea put forth by Macdonald and Pfitzer was eventually verified sixty years later by another German, Lothar Geitler.[28] Geitler tracked the size of cells in lab cultures over time and found they indeed progressively shrank. In principle, this could continue until a cell practically disappears. Does this actually happen? No, because the shrunken cells have a Houdini-like trick up their glassy sleeves: they are uniquely capable of meiosis. Haploid gametes are formed and released from the frustule in the presence of another compatible cell. These merge to form a diploid cell (the *auxospore*). The auxospore enjoys a brief period of unconstrained growth, after which a new frustule is built and the asexual cycles repeat.

In contrast to diatoms, other unicellular algae spend the vast majority of their time as haploids and immediately launch meiosis after a diploid cell is produced by syngamy. When starved of nitrogen, the haploid green alga *Chlamydomonas reinhardti* becomes capable of merging with other cells. There are two mating types, *a* and *α* (the Greek letter alpha), which are uniquely able to stick to each other. After fusion, the zygote soon launches into meiosis, producing four haploid progeny that remain in a tough cyst until the following spring.[29] The mating types are not the same as sexes (more on that in chapter 2). Rather, their function appears to be serving as a signal that the zygote is now diploid.[30] The mechanism by which this works is elegant: only diploid cells have both *a* and *α* mating type genes, and only cells with both types can produce a two-component factor that triggers meiosis. In this way, the change from haploid to diploid can be reliably recognized.

LINES AND CIRCLES

Eukaryotes generally have much bigger genomes than prokaryotes for two reasons: an increased number of genes as well as the proliferation of selfish DNA elements (often called junk DNA). Eukaryotes also package their genome's DNA into linear chromosomes rather than circles like their ancestors did. Because circles are endless, the presence of ends in DNA is seen as a sign of DNA damage in prokaryotes. The switch to linear chromosomes thus required new systems of "end management," indicating the change didn't happen casually. If circles had worked for billions of years, why bother? Linear chromosomes may have evolved simply because really huge circles are hard to replicate. But another likely stimulus for going linear is directly related to sex.

When linear chromosomes exchange pieces, the resulting recombinant molecules each have the complete set of genes in the correct order as long as they match their orientations. Now imagine an alternative universe in which only circular chromosomes exist in a cell undergoing meiosis (figure 1.6). Considering how exchange between such molecules might work, an immediate problem is where to start. With linear chromosomes, even a completely random system of pairing would get the correct alignment half the time. If correct pairing is more stable, the productive arrangement will be found quickly. But with circles, there are, in principle, almost infinite ways to get it wrong and only one way to get it right. This would complicate the pairing process substantially.

Assuming cells align their circular chromosomes correctly, we then encounter a second problem: one exchange (or crossover) between them ends up linking the two molecules into one double-sized circle (try this with two rubber bands to convince yourself that this is the case). When the first division of meiosis occurs, these circles would have to be either pulled apart imprecisely or passed to a single daughter cell, leaving the other without that chromosome altogether. The resulting imbalance of chromosomes is called *aneuploidy* and is usually lethal to a cell. Even worse, *any* odd number of crossovers has this same effect. Avoiding aneuploidy with circular chromosomes requires restricting crossovers to even numbers only (e.g., 2, 4, 6, etc.) for every chromosome to produce exactly one recombinant copy of the genome.[31] As genomes of eukaryotes grew, the multiple chromosomes likely selected for more reliable meiotic systems that were based on linear DNA molecules. The fact that linear chromosomes exist in eukaryotes as a direct facilitator of sex is further supported by the fact that some eukaryotes, such as fungi, still use circular molecules (called *plasmids*) for small, optional parts of their genome.[32]

Standard meiosis with linear chromosomes

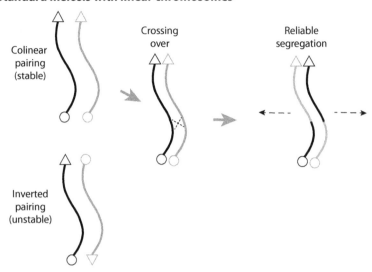

Meiosis with circular chromosomes

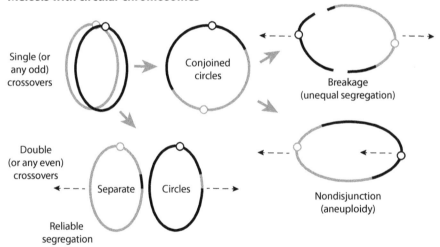

1.6 The advantage of linear chromosomes in meiosis. Top: meiotic recombination and chromosome segregation to daughter cells in eukaryotes, with linear chromosomes. Although a single crossover event is shown, any number produces two complete chromosomes. Bottom: hypothetical paths for meiosis with circular chromosomes. Even assuming correct pairing is achieved, reliable segregation requires an even number of crossovers.

Source: Redrawn from F. Ishikawa and T. Naito., "Why Do We Have Linear Chromosomes? A Matter of Adam and Eve," *Mutation Research* 434, no. 2 (June 1999): 99–107, https://doi.org/10.1016/s0921 -8777(99)00017-8.

LOOKING BEHIND, LOOKING AHEAD

In this whirlwind tour of the first few billion years of life on Earth, we have seen how the origin of sex links phenomena as huge as the global atmosphere and as tiny as mutations in the DNA of a single cell. An archaeon related to the present-day Loikiarchaeota adapted to dangerously high levels of oxygen by internalizing an aerobic bacterium. The resulting chimeric organism—the first eukaryote—was able to venture out of anaerobic refugia to which other Archaea were relegated. From this origin at a time of crisis, the widespread colonization of the Earth by eukaryotes began. However, their cells had to adapt to a changed internal environment as well. A new challenge was the elevated DNA damage from mitochondrial ROS. The sexual cycle, combining whole-genome recombination (meiosis) with syngamy, was a key innovation that allowed eukaryotes to thrive.

With sexual reproduction in place, the Eukarya diversified into myriad single-celled forms. Roughly 1 billion years ago, different groups of eukaryotes evolved bodies composed of many specialized cells that were permanently stuck together. Multicellularity introduced new challenges, complexities, and opportunities for eukaryotic life. As we will see in the next chapter, the combination of sex and multicellularity led to new phenomena that are at the heart of our human experience: mortality and the male-female distinction.

CHAPTER 2

Ex Unum Pluribus

Synopsis

Human fetuses begin with features of both male and female anatomy. Under direction of chromosomes and hormones, sex-specific bodies emerge by the selective loss of some features and the enhancement of others. Our pervasive sex differences evolved over hundreds of millions of years in response to challenges faced by our ancestors, and they are not easily interconverted after the fact.

———◆———

The following anatomical features are arranged in roughly the order in which they appear in human development:

blastomere
primitive streak
mesoderm
neural tube
somite
mesonephros
limb bud
gonadal primordium
Wolffian duct
genital tubercle
testis

glans penis
clitoris
labia majora
scrotum

If you are like most people, you found yourself snoozing through the first half of the list, woke up a bit at "genital tubercle," and perhaps began to wonder if anyone was looking over your shoulder at the end. This little exercise highlights how embryonic development gradually produces a sexualized body. What we are born with is further exaggerated at puberty. It is hard not to have opinions about the outcomes. From prude to libertine, burkas to bikinis, even nudists are making a decided statement when they declare that casual display of our male- and female-specific body parts is no big deal. Yet if we go back far enough, these salacious organs begin as very innocuous bumps and tubes common to both sexes.

Should you now be overwhelmed by an urge to study the development of mammals in three dimensions, an educational supply company will be happy to send you a preserved fetal pig. A really nice one, with the veins and arteries filled with blue and red latex, goes for about twenty-five bucks (plus shipping). The one I dissected in college is burned in my memory, both for what I learned about anatomy and for its unforgettable formaldehyde smell. Around a foot long, these specimens are developed enough that you can tell a male from a female with a little practice.

What if you want to appreciate the equivalent for our own species? Well, as you might imagine, that is considerably trickier. If you Google "human male female development" (like I did) you get lots of fairly informative diagrams, as well as plastic fetal replicas for sale (some wildly inaccurate). But what if you want the real deal? (Am I weird for wanting to see the real deal?) Given the hot-button nature of abortion, contraception, and fetal tissue research, how do scientists study human embryos and fetuses directly?[1] My search for answers to these questions eventually led me to a nondescript warehouse in the suburbs of Washington, DC.

If you pick up a textbook that describes human embryos, it will almost certainly refer to the Carnegie stages of development. Covering the first trimester of development, the stages were formalized by researchers at the Carnegie Institution of Washington's Department of Embryology. This august research center has operated in Baltimore, Maryland, since 1914.[2] Formed to manage and exploit a growing collection of human embryos gathered by its founding director, Franklin Paine Mall, it came to house hundreds of specimens. These came mostly from coroners and obstetricians, who were urged by Maryland's State Board of Public Health to send all fetuses to Mall. Many were carefully sliced into thin sections and prepared with stains, making them suitable for microscopic examination. Others remained intact, immersed in vats of fixative and awaiting a future mission. As nondestructive computed tomography (CT) and magnetic resonance imaging (MRI) matured, these embryos have found new importance.

I live and work twenty-five miles from the Carnegie, and as the content of this book became clear, I realized that their famed human embryo collection began just up Interstate 95. What luck, right? Wrong. By 1970, the mission of the Department of Embryology had become more focused on experimental biology. Today, one is far more likely to encounter a mutant fruit fly there than a pickled fetus. In 1973, they packed up the collection of human specimens that was their original raison d'être and sent it to the University of California, Davis (UC Davis). Like, on the other side of the continent. Sad trombones played in my head.

The itinerant embryos were not done travelling, however. After adding new materials related to bone growth, UC Davis relinquished the Carnegie embryo collection in 1990. Looking for a reliable entity to take custody of their "old-fashioned" embryo collection, they found a willing partner in the U.S. Army. Indeed, war is a sworn enemy of anatomical integrity (to put it nicely), and for this reason the army has a long-standing interest in the development (and possible regeneration) of the human form. The Carnegie collection was adopted by the Army-run National Museum of Health and Medicine, forming the core of what is now called the Human Developmental Anatomy Center (HDAC). As luck would have it, HDAC is based in Silver Spring, Maryland—even closer to my home and office than Baltimore. Twenty-three seconds after learning this fact, I emailed Elizabeth Lockett, the civilian scientist who manages the collection.

A month later, I found myself in an unusual situation for a biology professor: at the security gate of an army base, trying to talk my way in. After filling out a pile of paperwork and obtaining my official visitor pass, I was cleared to visit the world's greatest collection of human embryos and fetuses. It is housed in a remarkably unremarkable building—a windowless former food warehouse nestled among the auto body shops, plumbing supply companies, and pupuserias that make up a mid-Atlantic suburban industrial zone. I imagine this anonymity is no accident. I am advised to keep my phone and laptop Wi-Fi off, and to take no photos—a general aura of secrecy I'm not used to. Once inside, however, I find myself in a modern, brightly lit, and meticulously organized space. If human embryology has a great temple, this is it.

THE FIRST TRIMESTER

As we toured the collections, Lockett explained that "the Carnegie histology collection has eight hundred sets of completely sectioned embryos. These range in age from the first cell divisions through the end of the first trimester." With true professionalism, I replied "Wow."

"The oldest specimen required three thousand sections to get it all onto glass slides."

"Wow, wow."

In subsequent visits I learned more about the process, which represented what was literally cutting-edge science a century ago. A team of two researchers would carefully document newly acquired embryos with photographs and other information. A painstaking process of imbedding them in a solid support medium followed, after which each was carefully sectioned with a microtome. The microtome is essentially a miniature salami slicer capable of cutting as many as fifty slices per millimeter. Working in a team, Carnegie scientists would devote as much as three months to section a single embryo, mount the slices on glass slides, and reveal subtle differences in the miniscule bits of tissue with informative stains (figure 2.1). The resulting slides are simultaneously a form of data and little works of biological art, permanent testaments to both the plight of the mother from whom the embryos came and the dedicated people who made the most of each private tragedy. As I handle these one-hundred-year-old slides, I am awash in feelings of being fortunate and of acute reverence for my predecessors.

Sectioned embryos allow us to see the first events that distinguish male and female embryos from each other, which are completely internal and begin at about six weeks of development (figure 2.2).[3] Gonads are a mixture of germ cells and somatic cells. Germ cells are uniquely able to produce sperm or eggs. The somatic parts of the gonad support the germ cells. The remaining, non-gonad parts of the body are also considered somatic, and they collectively take care of the day-to-day business of living (more on that in chapter 3). In the first month of development, the germ cells complete a tricky migration from the gut to the site

2.1 The Carnegie Collection. One of the forty slides of the twenty-week-old Embryo 966 is at bottom. Each of the four sections is forty micrometers in thickness, about half that of a human hair.

Source: Elizabeth Lockett, Human Developmental Anatomy Center, National Museum of Medicine.

of gonad development, atop a temporary kidney called the mesonephros. After uniting with the somatic gonad cells, the resulting embryonic gonad is shaped like a tiny bratwurst and is at first identical in each sex. These primordial gonads form near a pair of somatic tubes, the Wolffian and Müllerian ducts. Between six and eight weeks, however, male and female gonads and associated ducts begin to pursue different paths.

In females, the first break from uniformity is the loss of a set of cord-like structures associated with the germ cells of the gonad. This isolates individual

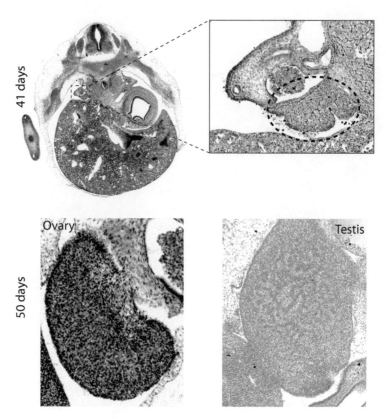

2.2 Divergence of human gonad development from a common starting point. At the top is a section of a forty-one-day (Carnegie stage 17) embryo, shown at low (left) and higher (right) magnification. The central nervous system and digestive organs are well developed, but the gonad is identical in both sexes. By fifty days (Carnegie stage 21, bottom), the testis (right) has developed the cords that will eventually form tubes of sperm-producing cells, while in the ovary (left), the cords are not seen, and oocyte follicles are starting to form.

Source: Specimens graciously provided by the Human Developmental Anatomy Center, National Museum of Health and Medicine, Silver Spring, Maryland. Photos taken by the author, with a Nikon E800 microscope, Q Imaging digital camera, and Image Pro Plus software. (*Top*) Carnegie Collection Embryo 6520, slide 43; (*bottom left*) Embryo 7392, slide 52; (*bottom right*) Embryo 7254, slide 52.

germ cells, like raisins in pudding, and foreshadows the eventual formation of a finite number of ovarian follicles, each with one egg. The developing ovary will retain the Müllerian duct, which will eventually become the oviducts and uterus. In males, the gonadal cords continue to grow, and the germ cells proliferate. This sets the stage for the eventual formation of the tubules of the testis, which contain both somatic Sertoli cells and millions of germ cells that are continually replaced throughout adult life. These will be connected to the Wolffian duct, which will form the epididymis (sperm storage organ) and the elongated vas deferens. The vas deferens carries sperm out of the body and is the target of the surgeon's scalpel in a vasectomy.

As we will see in chapter 4, the deciding factor that tips the human gonad into either male or female form is sex chromosome composition. The double dose of the X chromosome and the absence of a Y lead to the female pattern, while the presence of the Y and single dose of the X promotes the male.[4] But if chromosomes give the orders, then their lieutenants are hormones. In the seventh week of development, the future Sertoli cells begin to secrete anti-Müllerian hormone (AMH).[5] As its name implies, it promotes the degeneration of the Müllerian duct, leaving the Wolffian duct to serve the testis. Persistence of the Wolffian duct in males requires high levels of the steroid hormone testosterone, which is provided by the Leydig cells of the adjacent testes.[6]

The absence of Leydig cells in female embryonic gonads means testosterone is not produced by them, as it is in males. Ovaries also lack AMH-producing Sertoli cells. As a result, the Wolffian ducts degenerate and the Müllerian ducts persist. Further development of the Müllerian duct into the female reproductive tract is promoted by another class of steroids, the estrogens. In early development, they are made in the placenta rather than the embryonic ovary.[7] Although male embryos share this high-estrogen environment, their early sexual development is (necessarily) insensitive to it. Because testosterone is an intermediate on the path to producing estrogen in the placenta, female fetuses are also exposed to some "male" hormones. However, these levels are too low to trigger male differentiation. The overall result is that the internal plumbing that connects the gonad to the outside develop from truly sex-specific tissues. This lack of one-to-one correspondence between male and female is not seen in any other human organ system.

While the above events steadily push male and female embryos down different paths, they cannot be distinguished externally for another month or so, at the end of the first trimester of pregnancy.[8] Like the early gonads, the external genitals begin from a common starting point. Two months into development, both male and female forms have a little protrusion called the genital tubercle (figure 2.3). The tissue that surrounds its glans (or head) is called the prepuce. At around eleven weeks, the genital tubercle develops a groove on the lower side, a sort of trough that is connected to the tube that will carry urine from the

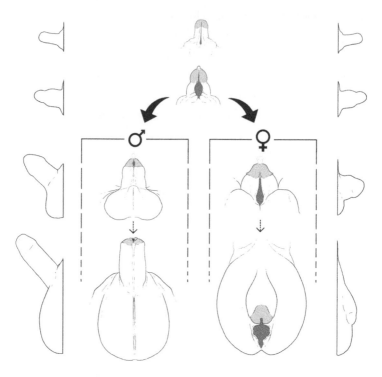

2.3 Developmental divergence of male and female external genitalia. The central panel depicts development of the genital tubercle (*top*) into the penis (*left*) or clitoris (*right*), with its tip, the glans, shaded to show how its position shifts in each. The urethral plate (groove just below the glans) opens in both sexes. In males, it "zips up," with the opening migrating to the tip of the penis; it remains open in females. The bulged structure below the genital tubercle is the labioscrotal swelling. Flanking diagrams are side profile views.
Source: Nicholas Bezio.

bladder, the urethra. In females this groove remains open, so that urine exits the urethra close to the body. In males, however, the edges of the folds that form the groove zip up to form a tube inside the future penis. In addition, the male prepuce extends all the way around the glans, causing it to protrude away from the body wall, whereas in females it remains mostly on the upper side. These differences in external genitalia (as well as the formation of the male-specific prostate gland) depend on the local conversion of testosterone to a more potent derivative, dihydrotestosterone, in males. They also ultimately explain one of life's greatest mysteries: why boys can pee standing up.

Below the developing genital tubercle is a pair of pouches, called the labio-scrotal swellings. In females these end up flanking the vagina as the labia majora ("large lips" in Latin). This occurs as the clitoris and the urethral groove settle down into it. In males, however, they form the future scrotum. After these external events have occurred, an ultrasound is usually able to reveal a fetus's sex.

SECOND AND THIRD TRIMESTER

Early in the second trimester, the basic setup of male and female reproductive systems is set, although in miniature. The next three months are not without drama, however (figure 2.4). In females, the posterior ends of the Müllerian ducts join, and then they begin to extend and thicken to form the precursor of the uterus and cervix.[9] The vagina itself, however, is formed from a rod of tissue that connects the cervix to a spot just below the developing clitoris (the urogenital sinus, which closes in males as the scrotum zips up). This means the inner and outer parts of the female reproductive tract form separately and need to form a common channel during development. How this happens has only recently been described.[10] There is initially no channel connecting the uterus to the vagina. Over time, however, one is opened by internal cell rearrangements. By twenty-two weeks (a little over halfway through full-term gestation), a tube connecting the ovary all the way to the vagina has formed. We all owe our births to that event in the early development of our mothers and in the many others before them.

Male fetuses undergo an equally momentous, if somewhat stranger, transition in the second trimester. If you have been paying attention, you may have noticed that human testes form in the abdomen, like ovaries, but they end up dangling between the legs, housed in the scrotum. The movement of the testes is a deeply weird feature of mammalian sexual development. Who would ever choose to have their family jewels in a place where they can get poked, snagged, or otherwise traumatized? Men around the world are generally united in our irritation at this arrangement, yet it is far from just a human thing. Testicular descent was already in place in the ancestor of all placental mammals, some 160 million years ago (see chapter 7).[11]

Several lines of evidence indicate that testicles are placed outside the body because spermatogenesis works better when it occurs below normal body

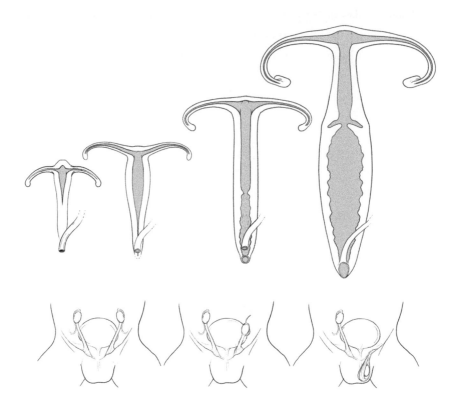

2.4 Later stages of internal sexual development of human female (*top*) and male (*bottom*). As shown at the left, the vagina is initially a solid rod below the uterus, with the urethra already connecting the bladder (*not shown*) to the exterior (see figure 2.3). Over time, a canal is formed that connects the uterus to the urethra in a single opening (*at right*). The constriction between the vagina and the uterus is the cervix. For males, the gonads (testes) begin in the same basic position as that of the ovaries, in the body cavity. During testicular descent, the gubernaculum contracts to pull them past the bladder and into the scrotum. Testicular descent is shown on one side only to illustrate the extent of movement.
Source: Nicholas Bezio.

temperature. For example, males whose testes fail to descend are sterile if the condition is not corrected by the time spermatogenesis begins at puberty. Otherwise, the testes atrophy and are prone to developing germ cell cancers. It is also true that prolonged soaking in hot water that is otherwise tolerable can temporarily reduce the sperm count of men.[12] Conversely, mammals that returned to life in cold water, such as whales and seals, lost the scrotum. Hydrodynamic streamlining may be an issue, but water is also an excellent heat conductor. Placing the testes near the skin surface, as in seals and whales, is adequate to cool them.

The "gotta keep the boys cool" idea is well supported, yet there are a few mammals that found a different solution. If you have ever seen a bull elephant's scrotum, you would presumably remember it given the size of male elephants. A bull elephant's scrotum is impossible to see, however, because elephants have neither a scrotum nor even descended testicles. And neither do their closest relatives, such as the elephant shrew, hyrax, and African golden moles. These mammals are descended from ancestors with a scrotum, but they have been making sperm just fine at core body temperature for millions of years. This suggests that once this innovation arose, it allowed loss of testicular descent and the obvious advantages that come with it.

With that digression aside, *how* do the testes end up in the scrotum in mammals like us? This is achieved by the action of a peculiar organ called the gubernaculum. This is a sort of stretchy ligament connects the testes to the scrotum. At about twenty-seven weeks of development, the gubernaculum begins to contract, pulling the testes inexorably down into the scrotum over a period of roughly two weeks. This journey is fraught with danger, and failure of descent is one of the most common birth defects in boys (over 1 percent at one year of age). The need for a route of passage through the body wall and into the scrotum (the inguinal canal) also creates an intrinsic weak spot in the abdomen. If tissues surrounding the canal become torn, loops of intestine can be pushed into the scrotum as well—the classic inguinal hernia that men will no doubt remember is tested by the "turn your head and cough" instruction (and a gloved hand) from the examining doctor. If we needed further proof that we evolved rather than being designed, I rest my case.

CORRESPONDING PARTS

Most body parts are decidedly not sexual (at least for most of us). From toes, tibias, and tonsils to pancreases, pituitary, and parotid glands, most of our body is there to keep us going day to day. These parts are far more numerous than those that define our sex. There is, however, a funny set of parts that are simultaneously sex-specific in function yet also present in each sex. Exhibit A: nipples.

They are pretty darned useless to men, yet we have them anyway. Why? The only plausible explanation is that, while women definitely have use for them should they nurse an infant, there is no real reason for a man not to have them. So the developmental circuits that produce them lack any sex-specific property, and everyone gets them.

But wait, there's more. The scrotum and its corresponding part in women, the labia majora, do quite different things, but in both sexes they are, shall we say, rather sensitive. Similarly, the process of penile erection has a counterpart in women: the swelling of labia minora (the "inner lips" of the vagina). Recall that the urethral groove that remains open in the female zips up almost completely in the developing penis. But zipped or not, the tissue forms a spongy structure called the corpora cavernosa (the "cavernous bodies") that can swell with blood upon arousal. In both sexes, these swellings position the glans of the penis and clitoris to maximize stimulation. This means that intercourse is actually uniting two diverged versions of the same embryonic structure. I was blown away when I first realized this, but it is not exactly a new idea: Aristotle envisioned something similar when he proposed that men and women are really just variants of a single sexual format.[13] What Aristotle did not know, however, was that the internal plumbing that connects the gonads to the rest of the reproductive tract are not similarly corresponding (see figure 2.4).

PUBERTY

The focus here has been on the development of sex differences that occur prior to birth. Of course, we all know that a host of additional changes will occur at puberty. Sex differences at puberty depend on steroid hormones produced by the gonads. While often described as being "testosterone in males, estrogen in females," this is a simplification that neglects important cross-sexed roles for each hormone.[14]

The production of the sex-related steroids is regulated in turn by nonsteroid brain hormones. These include the small glycoprotein hormones that regulate overall maturation, such as thyroid-stimulating hormone, and the gonadotropins, which specifically promote growth and maturation of the gonads. Boys' bodies are transformed through the disproportionate enlargement of the testes and penis, deepening of the voice, increase in muscle mass, and the growth of facial hair. Girls experience their own changes, including growth of the breasts, widening and curving of the hips, and the onset of menstruation (termed menarche; see chapter 9).

The changes involving gonads and genitalia are obviously essential for fertility and so can be seen to result from the most ordinary form of natural selection. However, the other traits that are elaborated at puberty have less obvious utility and are also inextricably part of sexual attraction. This indicates that they result from the process of sexual selection, which will be discussed in more detail in chapter 11. For our purposes in this chapter, our bodies have not only been shaped by selection for fertility but also by interactions between and within sexes with regard to mating and pair bonding. These tend to exaggerate the basic sex differences, which sets the stage for the gendered world in which we live. Some of the many consequences of this will be addressed in chapters 9, 10, 12, and 14.

PRACTICAL IMPLICATIONS

We have seen that, while male and female people began their embryonic lives with identical parts, by birth, major differences exist. To what extent are these reversible? Adult humans who adopt the hormonal profile of either biological sex develop some traits of that sex. These are the traits that normally appear at puberty. For example, XX people who elevate their testosterone levels into the range that is typical of males will grow facial hair, produce more muscle in response to exercise, and even exhibit male pattern baldness. XY people who suppress testosterone and increase estrogen will grow breasts. However, the early anatomical events of sexual differentiation cannot be changed by hormones, and after puberty, the gonads and external genitalia will remain as in the natal sex. As a result, surgery is required to change them.

Gender-affirming interventions, whether to address developmental anomalies or to support transpeople, are being improved all the time.[15] What is unlikely ever to become possible, however, is the conversion of a fertile individual of one sex to an equally fertile person of the sex to which they were not assigned at birth. This obstacle is imposed by the early, complementary degeneration of some of the other sex's parts. For example, without a Wolffian duct, females cannot be coaxed into producing a vas deferens. Similarly, at birth, males lack the Müllerian duct-derived tissue that forms the fallopian tubes, uterus, and cervix in women. Chapter 14 further examines some of the difficult choices that people may face in light of the above, but for the most part this subject is beyond the scope of this book.

As we will see in chapters 5, 7, and 8, our male and female anatomies first appeared as our ancestors moved onto land and evolved internal fertilization. In contrast, fish and amphibians, our vertebrate kin who continued to reproduce in water, generally lack external genitalia and use a common set of ducts to carry gametes to the exterior. As a result, even adult fish can be functionally sex-reversed simply through cross-sex hormone treatments.[16] One cannot fail to appreciate how different the experience of being human would be were this also true for us.

CONNECTING THE DOTS

In chapter 1, we saw how photosynthesis eventually gave rise to the origin of eukaryotes and, with it, sexual reproduction. Nearly 2 billion years later, *Homo sapiens* showed up. As the above summary of development intimates, we are highly complex multicellular animals, with hundreds of cell types and a division of labor between those cells that can undergo meiosis and form gametes and the much larger number of somatic cells that support them. With rare exceptions, our diverse cells are packaged into one of two alternative organismal forms (that is, sexes) by the various developmental processes reviewed here. These differences are at the root of some of society's biggest challenges. How did we get from microbes to mammals and from water to land? Why are male and female bodies different, and when and how did they get that way? And while we are at it, why aren't we all straight and cisgender? The remainder of this book is an effort to provide answers to these questions.

PART II

From Microbes to Mammals

CHAPTER 3

Go Big

Synopsis

Multicellularity enabled new ways of living, but it required specialization of cells for either reproductive or support roles. Both predictable mortality and the male-female distinction arise as direct consequences of this specialization.

Looking out the window of my home in suburban Maryland, I see big life. Oak, sweetgum, and tulip trees tower over the house, and deer and rabbits munch on my garden. Sex in these large, multicellular organisms is plenty apparent. The flowers and the songs of birds and tree frogs, which announce the arrival of spring, are about getting mates together. Ditto for the flashing of the fireflies (which are actually beetles). The maples dust the ground (and my car) with their pollen, and their seeds sprout in the roof gutters. It is hard to overstate how becoming big fundamentally altered the way in which organisms have sex. The shift from one- to many-celled organisms was the midwife that brought male and female into the world. It also brought death, at least in the way that we animals perceive it.

In chapter 1 we saw how a new kind of life, the eukaryotes, emerged in response to the Great Oxygenation Event and was forced to invent sex as a consequence. These early innovators were (and often still are) single-celled, tiny organisms. But besides their unique cellular components (like the nucleus and mitochondria), eukaryotes are also special because they alone gave rise to big life: the organisms whose bodies are formed from huge numbers of cells with highly specialized jobs. As with single-celled life, multicellularity began in the oceans and associated fresh water. But in time, most of the multicellular groups (the animals, plants, and fungi) invaded the land, and their reproductive strategies shifted in the process.

A POSTAGE STAMP SYNOPSIS OF MULTICELLULAR EVOLUTION

The oldest animals known are the Ediacarans. These oddly shaped creatures flourished for 30 million years before being replaced by the Cambrian Explosion of more familiar animals roughly 540 million years ago (MYA) (figure 3.1A–F). Smithsonian paleontologist Douglas Irwin has pointed out that the Ediacaran organisms actually overlapped with the oldest Cambrian forms,[1] but it is not clear how they are related. When researchers examine the DNA of current animals (all descendants of Cambrian-type creatures), they conclude that their last common ancestor lived at least 150 million years before their first traces were recorded in the rocks. The primordial animals inferred from these studies, if they really existed, were small and without hard body parts. But judging from the features common to their descendants, they must have had several different cell types and complex signaling systems that allowed them to coordinate their development and life activities.

Plants have their own evolutionary story. The likely forerunners of land plants were freshwater green algae called charophytes (figure 3.1G).[2] The earliest traces of land plants are spores that appear in the Ordovician Period (about 80 million years after the start of the Cambrian), with the oldest complete specimens from the Silurian (roughly 30 million years later).[3] By the early Devonian (420 MYA), land plants were abundant (figure 3.1H). Animals and the fungi specialized for decomposition could only make a living on land after plants had colonized it, and they soon did. However, fungi that form symbiotic colonies with unicellular algae (that is, lichens) may date back 635 million years (figure 3.1I).[4]

Less known than animals, plants, and fungi is a fourth group of multicellular forms that are often mistaken for plants, the large brown algae. They include kelp, which can make huge floating forests (figure 3.1J). Brown algae are photosynthetic, but this ability has a curious origin story. Their single-celled ancestors, which were not photosynthetic, internalized another entire eukaryote cell—a red alga. This process, called *secondary endosymbiosis*, turned a heterotroph into a new

.1 The major multicellular eukaryotes. (*A–C*) Early Ediacarans: (*A*) *Dickinsonia*; (*B*) *Yorgia*; (*C*) *Tribrachidium*. D) Mid-Cambrian trilobites from the Wheeler Shale of Utah. (*E*) *Sinoeocrinus*, a Cambrian echinoderm from China. (*F*) Cambrian annelid from the Burgess Shale of Canada. (*G*) *Nitella tenuissima*, a Charophyte alga thought o be closely related to land plants. The inset is a fossil egg case, or oocyst, of a related alga from France. (*H*) An arly fossil land plant, *Chaleuria cirrosa*, from the Lower Devonian of New Brunswick, Canada. (*I*) Six-hundred-nillion-year-old fossil lichen from Weng'an, China. (*J*) A forest of kelp off the coast of California.

Source: Reprinted from S. A. F. Darroch, E. F. Smith, M. Laflamme, and D. H. Erwin, "Ediacaran Extinction and Cambrian xplosion," *Trends in Ecology & Evolution* 33, no. 9 (September 2018): 653–663, https://doi.org/10.1016/j.tree.2018.06.003, used with ermission from Elsevier. (*D*) Photo by Michael Vanden Berg, used with permission of the Utah Geological Survey. (*E*) Jih-Pai Lin National Taiwan University). (*F*) Jean-Bernard Caron/Royal Ontario Museum. (*G*) Charles Delwiche/University of Maryland. *nset*) From J. T. Hannibal, N. A. Reser, J. A. Yeakley, T. A. Kalka, and V. Fusco, "Determining Provenance of Local and Imported Chert Millstones Using Fossils (Especially Charophyta, Fusulinina, and Brachiopoda): Examples from Ohio, U.S.A.," *PALAIOS* 28, o. 11 (2013): 739–754, https://doi.org/10.2110/palo.2013.110, used with permission of Joseph Hannibal and the Society for Sedimen-ary Geology. (*H*) James St. John (Ohio State University, Newark), via Wikimedia Commons, https://upload.wikimedia.org wikipedia/commons/5/59/Chaleuria_cirrosa_fossil_land_plant_%28Lower_Devonian%3B_New_Brunswick%2C_southeastern Canada%29_1_%2815518602081%29.jpg. (*I*) X. Yuan, S. Xiao, and T. N. Taylor. "Lichen-Like Symbiosis 600 Million Years Ago," *cience* 308, no. 5724 (May 13, 2005): 1017–1020, https://doi.org/10.1126/science.1111347, used with permission of the American ssociation for the Advancement of Science. J: Marty Snyderman and dtmag.com.

kind of autotroph. Only later did a subset of its descendants become big forms like kelp.

BIOLOGY MEETS GEOMETRY

Large organisms simply cannot function as proportionally enlarged versions of microbes. Take breathing as an informative case: in a single cell, no part of it is far from its surface. Gases involved in basic metabolism, like oxygen, can reach all interior points in a fraction of a second. As the absolute size gets bigger, however, the time required to diffuse increases disproportionately. For example, a spherical organism the size of a grape would need hours for the center to replenish the oxygen it uses if diffusion were the sole means of exchange. Worse yet, it's not just oxygen but nutrients and wastes whose exchange are also surface-area limited.

The size problem stems from the geometry you learned in middle school. The surface area of a given shape will increase as the *square* of the length, width, or other linear dimension, but the volume will increase as the *cube*. For example, doubling the radius of a sphere will quadruple its surface area but multiply its volume eightfold. The result is that the amount of surface area available to serve a given internal volume plummets for any shape as it gets larger (figure 3.2). Multicellular forms had to get around this basic fact in order to attain large size.

With only one or a few cell types available to make a multicellular body, there are two basic strategies. One would be to form a thin sheet, either flattened or in the form of a ball around a dead center. Growing in a narrow filament, perhaps in combination with branching, also allows all cells to remain near the surface. Growth in sheets and by branching are used frequently by simple colonial eukaryotes and some prokaryotes. Some algae form hollow balls, and many other algae, bacteria, and fungi are experts in filamentous growth. These forms are clearly successful, but a body built like a water balloon or constructed wholly of branched filaments can adopt only a few lifestyles. Animals generally have compact bodies that concentrate food for digestion in a stomach or its equivalent. Even sedentary filter feeders, like corals and clams, are big enough that they require very particular anatomies. Achieving them required the evolution of many specialized cell types, and this had an important impact on sex.

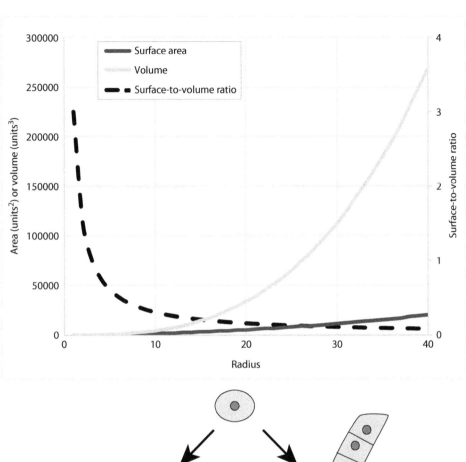

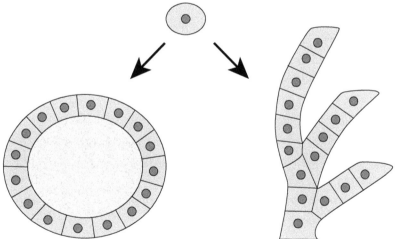

3.2 Growth and the surface area problem. Top: The impact of the radius of a sphere on its surface area, its volume, and the ratio of the two. The ratio plummets as the sphere gets bigger. Bottom: Two solutions for simple multicellular growth that maintain surface area. At left is a hollow ball of cells with a dead center (a flat sheet would achieve the same end). At right is a highly branched, filamentous form.

SEX AND THE SPECIALIZED CELL

In chapter 1, we saw that single-celled eukaryotes engage in sex with their whole (tiny) selves. Because big organisms depend on cell specialization, only some of their cells can be free to engage in the cycle of meiosis and syngamy that was the right of every cell of their single-celled ancestors. In animals, these special cells are called *germ cells*. The rest of the body's cells (collectively referred to as the *soma*) must perform the crucial support roles that allowed the organism to attain large size in the first place. This forced division of labor required multicellular organisms, however unconsciously, to decide how much of the body to allocate to reproduction.

A popular misconception is that the raw output of babies is the quantity that evolution maximizes. In truth, only offspring that survive to adulthood and can reproduce themselves (or to assist in the reproduction of their close relatives—see chapter 13) help their kind to persist in the long run.[5] Were an organism to commit 98 percent of itself to germ cells, it would have a big problem. Although it could, in principle, produce many embryos, in practice it would lack the wherewithal to gather the resources to provision and protect them. You might counter by proposing that if such an organism were very patient, and if it were to consume the necessary food slowly by some means or another, then it would work. The problem, however, is that such an organism would resemble a meatloaf: a highly nutritious blob that has no way to defend itself. Such organisms may have existed in the history of life, but presumably they all ended the same way: as some other creature's dinner.

The need to reproduce and survive simultaneously creates a trade-off for multicellular organisms: that which is invested in making babies cannot be put into maintaining the parent, or into protecting the babies once they arrive. Success in any particular habitat requires the right combination of investments. These include the obvious, such as in body parts that find and process the food required to produce big, fat eggs. But they also include behavioral investments, like the long-distance migration to breeding grounds undertaken by geese, salmon, and monarch butterflies. Once there, animals must also build nests or search and (in some cases) compete for a mate. In short, it takes a lot of soma to produce the next generation of squid, squirrels, or sea squirts.

As animals evolved, each lineage settled on a level of investment in gamete production that suited its lifestyle. This level varies greatly, yet in all cases, there is more soma than germ cells. Vertebrate gonads make up only 1 or 2 percent of the body,[6] and much of that is somatic support cells that nurture the gametes as they develop. At the other end of the spectrum, the nematode worms I study live in fleeting habitats packed with competitors. They must reproduce very quickly, and so they dedicate about a third of their adult bodies to their gonads, whose mass is almost completely composed of germ cells. With such an apparatus, one female can lay hundreds of eggs in a few days. And even that output is almost nothing compared to the parasitic nematode *Brugia malayi*, which causes a disease called lymphatic filariasis (or "elephantiasis"). To be transmitted from person to person, the mother worm floods the bloodstream of the person she infects with her larval babies. These are so tiny they can swim through our smallest capillaries until they are picked up by their ride: a female mosquito. One *Brugia* female lays over a thousand eggs per day, and she can continue this for years, adding up to billions of offspring over her lifetime.[7] But even in this extreme case, the worm is still over half soma.

Because of their highly specialized jobs, somatic cells are often unable to renew themselves. As they age, they become increasingly dysfunctional. Eventually, an essential component gives out, and the whole soma does too, often rather suddenly. Or maybe an annual plant finishes its seed production and soon freezes as winter sets in. These deaths are more or less predictable, but they are also just fine from an evolutionary perspective. As was first emphasized by the nineteenth-century German embryologist August Weismann, the germ line forms a continuous chain of living cells that connects all multicellular forms to the origin of life itself.[8] The germ line is thus, in a very real sense, potentially immortal. Alas, the soma that serves it need not (and typically cannot) be, and this is the sentient part of ourselves that we generally think of as our "self." Sperm and eggs carry the DNA and basic conditions to rebuild a new body from a single cell, but they cannot transmit learned skills or conscious memories. A central quid pro quo thus haunts our species: in exchange for evolving the anatomical and physiological sophistication that enables us to run, jump, invent calculus, write symphonies, and love, we must accept our eventual end. In short, the combination of multicellularity and sex gave birth to death as we experience it.

Nobody asked me if I wanted to be mortal—is there a way out here? What if my soma rebelled and reproduced outside the context of the germ line and gametes? In plants, fungi, and some animals, this can actually work. Through budding and regeneration, the soma can dodge mortality, sometimes indefinitely. In vertebrates, however, the equivalent process has another name: cancer. Cancer occurs when new mutations return highly specialized cells to a more embryonic-like, rapidly growing state. They stop taking orders from their surrounding comrades and go rogue. Early in animal evolution, surveillance systems evolved to block this process. These systems were generally advantageous because they protect the function of the whole organism and its sexual cycle. Given that their growing spree dooms the larger organism upon which they depend to survive, why does cancer persist? First, tumors have no foresight—they cannot consider their impact. But more important, cancer is mostly a disease of the aged. As we age, our somas matter less and less to the survival of our offspring, and eventually their failure is acceptable.

MALE AND FEMALE: THE SECOND GREAT QUID PRO QUO

Let's do a thought experiment: imagine a newly multicellular animal has evolved to spend 10 percent of its 1 gram (guppy-sized) adult body on gamete production, or 100 milligrams (mg). Let's also assume the two gametes that fuse to make the zygote are the same size. This is a reasonable starting point to consider because this condition (called *isogamy*) is found in nearly all unicellular eukaryotes. The question confronting our creature is this: how big should it make its gametes? Its options range from making one jumbo gamete of 100 mg to producing 100 gametes of 1 mg each (or even more that are even smaller). If all gametes produced equally viable offspring, the best strategy would clearly be to make them as small as possible so that their number is maximized. But for animals, there is another thing to consider: when the zygote forms, it is nowhere near functional as an organism. The chicken egg is a familiar example.

The yolk of a hen's egg is a single giant cell full of protein and fats, and it can't really do anything except be poached, scrambled, or fried. But over twenty-one days under the hen, it builds a body of roughly 1 trillion highly specialized cells, such as neurons and those that make up muscle, blood, cartilage, intestine, retina, gut, kidney, and so on. It finally hatches as a chick with a fully organized body that allows it to make its way in the environment. The need to produce a complex organism from a zygote introduced a new tension for multicellular eukaryotes: making gametes big enough to develop into a successful hatchling necessarily reduces their number.

While the development of healthy babies requires big gametes (or at least a large maternal investment in the case of species with pregnancy—see chapter 8), that is not the only consideration. Early animals were mostly sedentary (like corals and clams today), so achieving fertilization is not trivial. Besides being cheaper to make, smaller gametes also move better than larger ones, so we need to take that into account as well. The overall success is found by multiplying all the relevant factors together. For example, if parents make gametes that have a 50 percent chance of finding a mate, and the zygotes that form have 50 percent viability, then the overall success for each parent would be $(.50 \times .50) = .25$, or one quarter of the gametes they produce.

We could imagine that reducing a zygote's size by half reduces its viability by half but doubles its probability of finding a mate. These effects cancel each other out, and the best strategy is to make gametes as small as possible because this makes the most gametes. In nature, however, zygotes are typically the largest cells made by an animal during its life cycle. This tells us that large size is important.

The importance of being big suggests the following: beyond a healthy minimum (and fairly big) size, any further shrinkage triggers a disproportionate reduction in viability (figure 3.3). For example, cutting the size in half reduces viability by three quarters. The importance of this hypersensitivity was only recognized rather recently, first by British biologist Geoff Parker and his colleagues in 1972, and then by Canadian Graham Bell a few years later.[9]

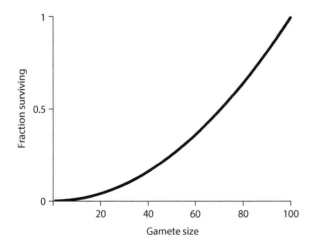

3.3 A plausible relationship between gamete size and zygote survival. In this model, reduction of gamete size inflicts disproportionate harm to the embryo.

Taking the mobility, postfertilization survival, and number of gametes that can be made at a given size into account, we see that there is one best isogamous gamete size. This lies at the value of half the maximal size (figure 3.4). In our example, that would be the 50 mg size, and our creature could make only two gametes. These are so big that many fail to reach a mate. As shown in figure 3.3, at that size, less than half of the zygotes that are successfully formed will produce surviving progeny. Between getting lost and dying from marginal resources, the majority of the gametes made are wasted.

There may have actually been a time when this compromise was the best early animals could achieve. As long as isogamy is the only option, any change in gamete size would make the overall outcome worse. Given this costly trade-off that wasted most of their germ cells, it should not surprise us that no present-day animals are isogamous. Instead, they all produce big eggs and tiny sperm, a condition referred to as *anisogamy*. Plants, kelp, and aquatic fungi are also anisogamous. The independent invention of sperm and eggs in very different organisms suggests it evolves as a direct consequence of multicellularity. The above thought experiment gives us a framework to help us understand why.

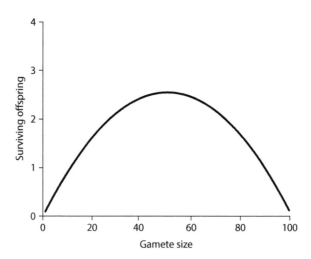

3.4 The isogamy compromise When gametes must both move and contribute half of the material needed for embryo survival, the best strategy is achieved at an intermediate size. Even at this optimal size, only a small fraction of gametes leads to viable progeny. In this example, with 1,000 mm³ of material for gamete production, only two to three of the twenty gametes produced survive.

The job of sperm is to guarantee fertilization and initiation of embryonic life and has little or no role in the subsequent viability of the offspring. In the context of our model, a male represents a motility specialist who can assume that the materials needed for zygote survival will be provided by the egg. In such a situation, sperm are selected to be small and numerous. Conversely, females are viability specialists. They can largely ignore the task of getting the gametes together but take on the entire burden of provisioning the zygote.

If we input the same total material to make gametes in both the isogamy and anisogamy scenarios, we find that both egg makers and sperm makers produce more successful offspring than isogamous alternatives. For example, while the isogamous scenario in figure 3.4 produces 2.5 viable offspring, the same material yields ten with male-female specialization. This is a huge difference that strongly selects for anisogamy. The takeaway here is that anisogamy evolves because each sex can count on the other to enable its role—a quid pro quo that is mutually beneficial.

While this thought experiment is plausible, it could still be completely wrong. Fortunately, evolution has provided us with a handy test of our model, in the form of fungi. Like all multicellular life, fungi originated in the water, and aquatic fungi called chytrids remain today. Chytrid life cycles are remarkably plantlike, with an alternation of diploid and haploid multicellular phases. The haploid stages produce distinct male (small) and female (large) gametes. Using their flagella, the sperm locate the larger eggs by chemoattraction. The size difference between male and female chytrid gametes is not as extreme as in animals and plants, probably because the zygote only needs to produce a very simple baby fungus before it is able to grow on its own. Still, this tells us anisogamy was a feature of the aquatic ancestor of all fungi.

When fungi invaded the land, they continued their original way of feeding. They digest organic matter with secretions from their filamentous bodies and then absorb the nutrients that are released to fuel their growth. In this way, the branched bodies of fungi function like one big stomach-intestine combination. As plants colonized land, they created abundant food for fungi, who followed them ashore. This opened a new realm for fungi, but the movement to dry land deprived them of the liquid medium their ancestors used to get their swimming gametes together. Terrestrial fungi responded by evolving gametes and zygotes that remain attached to the parent's bodies.

Once formed, the fungal zygote produces a new organism, often still connected to its haploid parents. Eventually the new diploid organism gives rise to a
mushroom or other fruiting body. It is in this body that the cells undergo meiosis, and new haploid cells (the spores) are produced. Comparing this cycle to that
of animals, the lack of both gamete motility and of an independent embryonic
stage that requires provisioning jump out. This means that both incentives that
promote anisogamy are lacking in terrestrial fungi. Remarkably, they are also the
only group of multicellular organisms that are isogamous. I rest my case.

For biologists, the production of either sperm or eggs is the defining feature
of male and female, respectively. But the presence of two gamete types doesn't
require separate male and female sexes. Many species of animals (and nearly all
plants) are *hermaphrodites*, defined as producing both sperm and eggs. Hermaphroditism is not synonymous with asexuality. Although hermaphrodites produce
both eggs and sperm, most still mate with another individual most or all of the
time. Of course, many other animals (including the birds and the bees) have the
familiar separation of male and female into separate sexes. There are even animals in which a pure male or female sex coexists with a hermaphrodite sex. All
these are sexual systems but with different packaging of sperm and egg production into bodies. Why particular sexual systems are used in animals and plants
will be tackled in the next chapter.

CHAPTER 4

The Left Fin of Darkness

Synopsis

Distinct male and female bodies seem normal for us humans, but the alternative arrangement of a single hermaphroditic sex is common in animals. Hermaphroditism is not restricted to squishy invertebrates, and it is not always ancient. Fish provide fascinating examples of sexual fluidity, and they also provide clues about why and how it evolves.

Sometimes a research article reports something we weren't even looking for before it was discovered. One of my favorites from the last decade is "Disposable Penis and Its Replenishment in a Simultaneous Hermaphrodite," by Ayami Sekizawa and her colleagues.[1] They studied the sea slug *Chromodoris reticulata*, which has one hermaphrodite sex. When the time is right, they pair up and exchange sperm by each extending a transparent penis. Afterward, the partners pull away while their penises are still connected, stretching them like salt water taffy. The mates eventually disengage but shortly thereafter proceed to lop off their own now-elongated members, discarding them on the sea floor. Fortunately, the lost organs soon regenerate, and the next day they are ready to go at it again.

C. reticulata may have strange disregard for its private parts, but it is not unusual in being a hermaphrodite that mates. Earthworms, slugs, and nearly all flowering plants make babies this way. So do flatworms, and they rival the nudibranchs in the oddity department. Many transfer sperm by puncturing the skin of the mate with a hypodermic penis. For obvious reasons, the worm would

4.1 Penis fencing in hermaphroditic *Pseudoceros bifurcus* flatworms.
Source: Photo by Nico Michiels.

prefer to be the stabbee only enough times to fertilize its eggs—usually once. Being the stabber is not so bad, however, and would potentially increase off-spring number. This dynamic in one species has led to a behavior called penis fencing, in which animals attempt to inject their mates without being insemi-nated themselves (figure 4.1).[2]

Penis-fencing hermaphrodites are not what most people think of when they are asked to imagine animal mating. More likely they would think of rams head-butting in an open meadow while ewes look on or a monogamous pair of pen-guins tending dutifully to their fluffy chick. This male-female version of sex is familiar in part because we do it that way, and many other animals share the separation of sexes with us. Were the first animals hermaphroditic, or did they have separate sexes? How common is it to change from one to the other?

To answer these questions, it would be nice to have a time machine that could take us back to the Cambrian Period (or before). Lacking that, we must settle for more indirect evidence. Fossils can sometimes preserve evidence of distinct

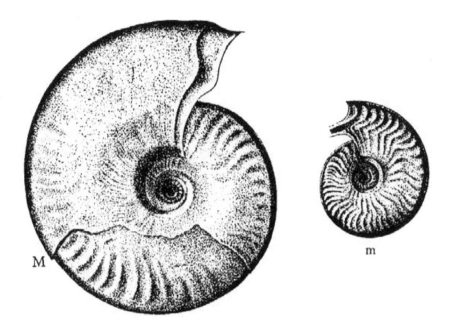

4.2 Sexual dimorphism in Jurassic ammonites. The larger form (M) was ascribed to the genus *Graphoceras*, and co-occurs with the smaller form (m), originally described as *Ludwigina*. Based on details of growth and their consistent co-occurrence, J. H. Collomon proposed these were female (M) and male (m) forms of one species.

Source: Reprinted from J. H. Callomon, "Sexual Dimorphism in Jurassic Ammonites," *Transactions of the Leicester Literary and Philosophical Society* 57 (1963), used with permission of the Society.

male and female forms (figure 4.2), but such cases are rare. Even when males and females differ greatly, they can be recognized as sexes only when the two forms are found consistently together in multiple places. In rare species or those in which males and female occupy different habitats most of the time, the sexes can be mistaken for two distinct species.[3] That leaves an even more indirect approach based on the relationships of living organisms. Biologists visualize these relationships with phylogenetic trees, which since the 1980s are mostly produced by comparing the sequences of genes that are shared by all animals. The more similar the sequences, the more recent the split that separated the creatures that bear them is inferred to have happened.

An evolutionary tree showing the relationships of the major animal groups (called *phyla*) is shown in figure 4.3. If we note the sexual system used by each

4.3 Reconstructing the evolutionary history of sexes in the animals. The relationships of the animal phyla with more than a handful of species, with relatedness indicated by the total length of the path connecting them. The Bilaterians (boxed in gray) are named for their bilateral (left-right) symmetry. The most likely ancestral sexual system inferred from existing species is indicated above each phylum, with "H" for hermaphrodite and "MF" for male-female. "MF*" indicates that, while the ancestor was male-female, a large proportion of living species are hermaphroditic. Bryozoans are labeled "?" because they are hard to categorize (see text). Ovals mark two alternative reconstructions of the evolution of separate male and female sexes. Darker ovals mark a single origin in the ancestor of cnidarians and Bilaterians (solid black), with reversion to hermaphroditism (dark barred ovals) in the acoels, arrow worms, and flatworms. Lighter ovals depict two independent origins in the cnidarians and at the base of the Bilaterians (solid white), with reversion in arrow worms and flatworms (light ovals with thin bars). The gray oval depicts the switch to colonial growth.

Source: Phylogeny adapted from A. Hejnol, M. Obst, A. Stamatakis, M. Ott, G. W. Rouse, G. D. Edge-combe, P. Martinez, et al., "Assessing the Root of Bilaterian Animals with Scalable Phylogenomic Methods." Proceedings of the Royal Society B 276, no. 1677 (December 22, 2009): 4261–4270, https://doi.org/10.1098/rspb.2009.0896; K. M. Kocot, "On 20 Years of Lophotrochozoa," Organisms Diversity & Evolution 16 (2016): 329–343, https://doi.org/10.1007/s13127-015-0261-3; F. Marletaz, K. Peijnenburg, T. Goto, N. Satoh, and D. S. Rokhsar, "A New Spiralian Phylogeny Places the Enigmatic Arrow Worms Among Gnathiferans," Current Biology 29, no. 2 (January 21, 2019): 312–318, https://doi.org/10.1016/j.cub.2018.11.042; P. Simion, H. Philippe, D. Baurain, M. Jager, D. J. Richter, A. Di Franco, B. Roure, et al., "A Large and Consistent Phylogenomic Dataset Supports Sponges as the Sister Group to All Other Animals," Current Biology 27, no. 7 (April 3, 2017): 958–967, https://doi.org/10.1016/j.cub.2017.02.031.

group, we can use the tree to examine alternative scenarios and recognize the most likely one. The two lineages that split off from the rest of animals first are the sponges and ctenophores (comb jellies). Today, nearly all of them are hermaphrodites.[4] While it's possible that they independently switched from an ancestral state of separate sexes to become hermaphrodites, the simpler inference is that they have been this way from the start. If the latter is true, then the ancestor of all animals was also probably hermaphroditic.

The remaining phyla are a mix, and some have both hermaphrodites and male-female species within them. For example, among cnidarians, most reef corals are hermaphrodites, but other corals, anemones, and jellyfish are male-female. But more focused analysis of many cnidarian species indicates that the ancestor of the entire group was most likely male-female.[5] The same is likely true for annelid worms; although the familiar earthworms are hermaphrodites, their marine annelid ancestors were male-female.[6] By carefully examining each phylum in this way, I conclude that all but four clearly had the male-female system as their ancestral state. One of these exceptions, the bryozoa, is colonial and often separates male and female into different subindividuals. It is plausible that bryozoans descend from a solitary ancestor that was male or female, but the history is admittedly ambiguous. The other three exceptions, the acoel worms, arrow worms, and flatworms, have likely been hermaphroditic since they split off from other animals. The simplest way to account for their presence is as reversals from an ancient male-female ancestor.

Amid the many details, the reconstruction in figure 4.1 has a rather startling implication: *the existence of separate males and females in our species can be traced back to (or before) the Cambrian.* So that twinge of kinship you may feel when you see a mother spider tending her egg case is not misplaced—motherhood in its purely female form descends from an ancient progenitor shared by you and the spider. I am personally thankful, however, that the female spider's tendency to eat her mate during or shortly after insemination is not so universal.

WHY CHANGE?

The evolution of separate male and female sexes is without a doubt one of the more important events in the history of life, one that affects us in myriad ways to this day. But why did it happen? If animals were initially successful as hermaphrodites, and often still are, why change? This question has been pondered by many smart people,[7] but fortunately the general form of the answer is simple. As we did for the origin of the sperm-egg distinction (anisogamy), we can use a simple thought experiment to understand the basics.

Imagine a population of hermaphrodites that mate, as earthworms do today, by trading sperm. Each makes ten eggs in its ovaries and enough sperm in its testes to fertilize ten others. The result is that every individual is the parent of twenty offspring, ten as the mother (via eggs) and ten as the father (via sperm). As we think about how an alternative, more specialized strategy might take over, the key insight is that to do so, it would need to produce more than twenty total offspring given the same input of resources.

Into our balanced hermaphrodite population, we can now insert a new sort of hermaphrodite, one that invests asymmetrically in one sex's function. For example, it might make more eggs but fewer sperm, or deliver more sperm but make fewer eggs than before. For this to spread, the increase in investment in one sex must more than make up for the loss of investment in the other. In the example above, more than twenty offspring would need to be produced.

One such strategy could be a sperm specialist—a true male. Sperm are tiny, and many can be made for the same resources required to make a single egg. For species that release gametes into the water, a dedicated male that doesn't have to hold resources for egg production may be able to make so many more sperm that he sires a large proportion of the offspring in the population. Similarly, for animals that mate by direct contact, perhaps a dedicated male could move quickly from mate to mate, inseminating many hermaphrodites before he runs out of ammo. Both sperm-only strategies should quickly spread if they can more than double the males' offspring. They cannot take over completely, however, because if all individuals adopted it, there would be no eggs to fertilize.

How does a true female evolve? The reliable presence of males allows females to stop making their own sperm, but that change alone would produce fewer babies, not more (the ten eggs alone versus twenty total in our hypothetical hermaphrodite). As with males, to gain traction, a dedicated female would have to make more than twice as many eggs as her hermaphroditic ancestor. Because sperm are tiny, it would at first seem the reallocation of these resources would fall short of what is needed to more than double egg production. However, we must factor in the other cost associated with male functions in hermaphrodites. Elimination of all male structures (like the harpoon penises of flatworms and the muscles required to launch them) may free up anatomical resources. Fertilizing another individual's eggs can also require spending considerable energy searching for and courting mates. This can exceed what is required to make sperm. Females could stop these activities and perhaps increase egg production enough to make it a win for them as well. The conditions required to switch sexual mode would thus be peculiar to an animal's lifestyle, and this probably explains why there is variety today.

WHY CHANGE BACK?

We just saw some circumstances in which male and female sexes can evolve in hermaphroditic animals. However, there are also many cases where species or groups of related species have abandoned separate sexes to return to hermaphroditism. For example, a subset of mollusks and annelids, the flatworms, barnacles (which are actually a weird sort of crustacean), and most reef-building corals are hermaphrodites, even though figure 4.1 implies their ancestors were male-female. If having separate sexes is so great, why change back? The general answer must be the converse of the argument for why separate sexes evolve: something about being hermaphroditic increases reproductive success over what it would be for the male-female situation.

One possibility is that difficulties in finding a mate makes having separate sexes risky. Imagine a male-female animal that requires a habitat that is rare and patchy. In such a case, new arrivals to a spot may find they are the only one of their kind present. No mate, no babies. Hermaphrodites make both sperm and eggs, and it may not take much of a tweak for them to combine them. This is referred to as "selfing" and is found in some hermaphroditic plants and animals.[8] Even if most reproduction is through mating, selfing is a valuable insurance policy. Sometimes it hurts, however: flatworms that inject sperm into their mate have been observed to self-fertilize by impaling themselves.[9]

Self-wounding aside, the more general downside to selfing comes from mutations. Outcrossing often prevents mutations that damage a gene's activity from causing harm by pairing them with a good copy in the offspring. As long as two individuals have different bad mutations (which they will if they are unrelated), most remain harmless. On the other hand, mating with a close relative, and especially with oneself, often pairs two bad copies, leading to loss of the gene's function and potentially sickness or death. This phenomenon is referred to as *inbreeding depression* and would be expected in any animal that tries selfing after eons of enforced outcrossing. We humans presumably eschew incest for this reason.

To get a sense of how bad inbreeding depression can be, consider a situation in which a hermaphrodite parent carries one broken copy of an important gene and one normal copy (i.e., they are a *heterozygote*). They themselves are fine, but there is a 50–50 chance any one of their sperm or eggs will carry the mutation. Because the chance of two independent events occurring is the product of their individual probabilities, one time in four a self-fertilized offspring would be a double mutant (or a *homozygote*). This might seem unlikely enough to roll the dice, but researchers at the University of Washington have estimated that the average human carries nine hundred harmful mutations in their genome.[10] We would thus expect over two hundred genetic train wrecks to result if a person were to self. If you take anything away from this book, it is this: *don't mate with your yourself.*

Although inbreeding can get ugly, it still might not stop the reemergence of a self-fertile hermaphrodite sex if mates become rare enough. Even with massive mortality, the reproduction achieved by selfing is still an improvement over nothing. In the longer term, the genetic carnage that is inbreeding depression can also have a cryptic upside: it tends to eliminate the worst mutations from the species. Over time, the least-mutated genotypes come to dominate the population, a process referred to by geneticists as *purging*. The "happy homozygotes" that result from purging are indeed observed in species that have mostly selfed for many thousands of generations. These include two of the most studied organisms in biology, the mustard weed *Arabidopsis thaliana* and the nematode *Caenorhabditis elegans* (which my own laboratory studies). Both species colonize small, isolated, and temporary habitats—exactly the situation we expect to promote selfing. That *C. elegans* went through a purging phase in the past can be confirmed by forcing closely related male-female species to undergo brother-sister mating. Although they have no ethical qualms about it, their offspring get progressively sicker, typically until the population goes extinct.[11]

A bit closer to us on the tree of life, there is one vertebrate animal that routinely self-fertilizes. Were you to visit the coast of South Florida, you could encounter the mangrove killifish, *Kryptolebias marmoratus*. At first glance, you might conclude that they were nearly all female, as most are drab and lay eggs. However, were you to follow a "female" fish from birth to adulthood, you would see that no mating is required for her to produce fertilized eggs. *K. marmoratus* achieves this through a subtle form of hermaphroditism: they possess small bits of testis tissue in what is otherwise a pretty normal ovary.[12] Sperm and eggs mix before laying, so that nearly all are fertilized by the same animal. Consistent with selfing as a tool of colonizers, *K. marmoratus* is the only member of its genus to have spread into the Caribbean from its ancestral home in coastal Brazil. This was aided as well by their unusual tolerance of variable water salinity and by their babies' habit of spending a month or more in the egg shell after they are fully developed. Add a few hurricanes to this, and soon there are *K. marmoratus* on every island.

A second circumstance that seems to favor hermaphroditism is the brooding of babies inside the parent. This resembles pregnancy but without transfer of food from mother to offspring after fertilization of the egg. Brooding is most common in small animals whose ancestors were larger ones that spawned their eggs into the water. At some point, as they got smaller, this strategy became unsustainable: too few gametes were made to guarantee even one success. By self-fertilizing and then holding formerly mobile larvae until they are safely past the point of metamorphosis, mortality is greatly reduced.

One of my favorite examples of brooding is the itsy bitsy, teenie weenie, yellow sea star, *Parvulastra parvivipara*, whose reproduction was first described by my Australian colleague, Maria Byrne. Adults are 5 mm across, the width of a pencil (figure 4.4). Hermaphrodite mothers fertilize and brood their eggs until

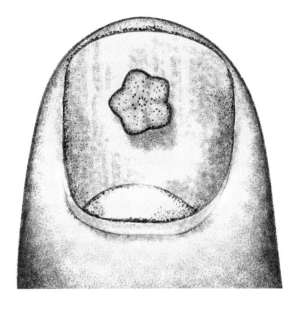

4.4 The tiny hermaphroditic sea star *Parvulastra parvivipara.*

Source: Drawing by Michiko Hsieh Haag, inspired by a photo by Flikr contributor Nuytsia@Tas (https://flic.kr/p/4xfnfk).

they metamorphose, and they even allow them to graze on their own tissue until they feel ready to crawl out and make their way in the world. Who knew cannibalism could be so cute?

Given the tiny number of eggs produced by *P. parvivipara*, releasing them into the water would rarely lead to any survivors. At the same time, males using traditional spawning would find that the tiny amount of sperm they could produce would be inadequate. While contact mating could solve the latter problem, becoming a self-fertile hermaphrodite was apparently the quicker route to reliable reproduction.

Barnacles are also hermaphrodites that brood their young. We know a lot about them in no small part because of Charles Darwin. While he was refining his case for evolution through natural selection, Darwin spent eight years standardizing the names of their various body parts, meticulously describing their many species, and using them as a case study to apply evolutionary thinking to how their diverse forms came to be.[13] More germane to our discussion here, he also discovered that, while most species are purely hermaphroditic, some also produce tiny males that settle within their hermaphrodite mates, and that in others the hermaphrodites are replaced by true females. The resulting tomes on this work, published between 1851 and 1855, solidified his credibility as a naturalist.[14] They also helped ensure that his 1859 bombshell, *On the Origin of Species by Means of Natural Selection*, was taken seriously.[15] So it is not untrue to say that barnacle sex helped change the world.

Unlike the hermaphroditic sea stars, barnacles are prodigious maters, transferring sperm directly to any mate within reach of their really long penises. Darwin himself was impressed, describing them as "prosciformed and capable of the most varied movements" (that is, like a highly maneuverable elephant's trunk). There is also evidence that sperm can be transferred to mates beyond their immediate reach. Unlike the tiny sea stars, it seems barnacles are really good at mating.

If inability to mate isn't the issue, then why did the barnacles take up hermaphroditism? One possibility is that, in every cluster of barnacles, someone has to arrive first. Perhaps the main purpose of hermaphroditism is to enable a lonely pioneer to reproduce solo until it attracts more neighbors to join it. Consistent with this, barnacles isolated in a laboratory can self-fertilize.[16] Another possibility is that even when attached to a mate, sometimes the small dedicated males can't compete with the male aspect of much larger hermaphrodites and are pushed out of the species. If true females rely on males, hermaphroditism is stabilized. In any case, because these different sex strategies coexist in similar habitats, it appears that nearly two hundred years after Darwin first pondered them, barnacles are still mysterious and still have plenty to teach us.

SEX AND SCIENCE FISHIN'

In her classic 1968 novel, *The Left Hand of Darkness*, Ursula K. Le Guin imagined an icy world called Gethen and its human-like organisms. In a new introduction to the novel written years later, she explained that she wrote the novel as a lengthy answer to this question: "If warfare is predominantly a male behavior, and if people are either male or female for only a few days a month during which their sexual drive is overwhelmingly strong, will they make war?"

To conduct this literary thought experiment, Le Guin imagined the Gethenians as hermaphrodites that can act as either a gestating mother or a sperm donor father, depending on the reproductive cycle. In addition, they are decidedly uninterested in sexual activity outside *kemmer*, a brief estrous period. They find their loss of control during this time rather embarrassing and generally go into hiding with their mate until it passes.

Humans have sexes that are both fixed and plural, and societal norms have accumulated around them. That Gethenians lack both leads to important distinctions between their society and our own. In the prologue, Investigator Oppong (a kind of cultural anthropologist/scout) notes the absence of things humans experience. Children have no distinct psychosexual relationships with each parent. The mechanics of *kemmer*, which depend on mutual arousal, eliminate rape and lack of consent. The lack of distinct sexes means there is no division into stronger/weaker, protective/protected, dominant/submissive. This all seems like a utopia, but Oppong points out that it also creates challenging cultural dissonances when Gethenians and humans interact:

> The First Mobile [envoy] must be warned that unless he is very self-assured, or senile, his pride will suffer. A man wants his virility regarded, a woman wants her femininity appreciated, however indirect and subtle the indications of regard and appreciation. On Winter [the name humans gave the planet] they

will not exist. One is respected and judged only as a human being. It is an appalling experience.[17]

I'm a big fan of Le Guin's novel, and a few years ago I had a chance to discuss the biology it imagines for the annual conference of the Baltimore Science Fiction Society. The goal was to identify earthly counterparts for different features of Gethenian sexuality. The strict confinement of sex to a brief estrous is common in many mammals, although not our closest primate relatives. Female dogs are famous for their brief, intense "heats" that attract males from all around the neighborhood, for example. So that idea is hardly far-fetched. (The lack of human estrous will be discussed in chapters 9, 10, and 11.)

With regard to having a single hermaphrodite sex, we saw above that this is also fairly common. Even a small fraction of our fellow vertebrates, the bony fish, produce simultaneous hermaphrodites with both ovary and testis tissue. Among these the prize for Most Gethenian goes to the tobaccofish (*Serranus tabacarius*). This Caribbean reef dweller alternates between providing eggs and donating sperm in successive rounds of mating with the same mate.[18] This is the closest approximation to Gethenian sex I know, but even tobaccofish do not combine sexual alternation with pregnancy. They evolved their strictly hermaphroditic reproduction rather recently, yet they are not tiny (like barnacles or miniature sea stars) or known to self-fertilize. The forces promoting this novel system remain unclear, but it has been proposed that the requirement for sperm donors also to offer eggs is a way to prevent fish from monopolizing multiple egg producers and reproducing solely via sperm.[19] Such a one male, many female (*polygynous*) system was likely used by the tobaccofish's ancestors.

LOOKING AHEAD

In this chapter, we have seen that animals vary a great deal in how they package male and female parts into their bodies. From a hermaphroditic beginning, most adopted specialization: separate male and female versions of their kind. Some reverted to a single hermaphroditic sex, however, and this occurred in many different groups. By comparing related animals with different strategies, we can start to see why changes occur, but much of this diversity remains utterly mysterious.

From this point forward, we will focus on organisms with separate male and female sexes. This is partly so we can lay the foundation for understanding our favorite species: *Homo sapiens*. But before we get there, it is worth exploring how one species can produce two distinct forms, one male and one female, from instructions encoded in the same (or nearly the same) genome. This is the general problem of animal sex determination and is the subject of the next chapter.

CHAPTER 5

One Genome, Two Bodies

Synopsis

In animals with two sexes, something dictates whether the body is male or female. That something is often a genetic difference, but age and environment can also serve. By 1990, the genetic circuits that regulate sexual fate had been identified in fruit flies, mammals, and nematodes. But there was no correspondence between any of them. When a common gene was finally identified, it resolved an enigma and confirmed the single origin of male and female in the animals.

———————— ◆ ————————

In chapter 4, we saw how the distinct female and male sexes of most animals evolved from hermaphrodites that had both male and female reproductive capacities. Forming separate sexes is as remarkable as it is familiar. Consider that identical (or nearly identical) genomes nevertheless produce two distinct, often very different animals. Such a binary outcome is necessary: the two sexes perform precisely complementary roles, and intermediate forms are generally not fertile. Because of strong selective pressure on reproduction, every animal with two sexes has a system of self-reinforcing signals that guides the process of development, from a common starting point, down the female or male path. This is referred to as the *sex determination pathway.*

When these sex-determining signals derive from alternative states of the genome set at fertilization, it is referred to as genetic sex determination (GSD). This often involves special sex chromosomes, like our X and Y. There are three common GSD options that produce equal numbers of males and females. One is the XX female, XY male system, found in flies, humans, and many other species. Note that the gene content of the X and Y chromosomes are not the same in all animals that use this system. Instead, "XY" simply denotes a GSD system in which the male has two different sex chromosomes, while the female has two copies of the same one. Because the male produces two types of sperm (X-bearing and Y-bearing), males determine the sex of offspring.[1]

Another common GSD system, XX-XO, is a variant of the first; males lack a Y, but still determine the sex of the offspring because half of their sperm carry an X and half do not. Finally, the female could be the sex to bear two different sex chromosomes (called W and Z to distinguish them from the XY system), while males have two Z chromosomes. Birds and butterflies are familiar examples of ZZ-ZW animals, but they are also found in some snakes, fish, and other animals.

The XY, XO, and ZW GSD systems all ensure that there is a clear starting difference between male and female embryos, and they also lead to equal numbers of males and females (with a few interesting exceptions). Bees and wasps, however, have a distinct form of GSD that does not guarantee an equal number of males and females. Here, the unfertilized haploid eggs of the female (the queen in social species) develop exclusively into males. The eggs that are fertilized, however, develop into females. Because the female can control whether she fertilizes her egg or not, the sex ratio can deviate greatly from the familiar 1:1 ratio.

Sometimes local conditions in which an animal develops dictates sex instead of a chromosome. This is referred to as environmental sex determination (ESD), which varies greatly in how it is regulated. In some cases, the key environmental variable is the presence of other individuals. For example, the marine annelid worm *Bonelia viridis* spends its adult life in the sediment, rarely moving. The more mobile larval worms that pioneer a new patch of habitat invariably develop as females. When another larva comes along, however, it will sense the female's presence, settle inside her, and become a diminutive sperm donor. For turtles and alligators, the temperature of the nest in which the eggs develop is the key factor. In these systems, the ratio of males to females could diverge greatly from 1:1.

While GSD and ESD would seem to cover the options for sex determination, there is a third phenomenon that merits mention here. In animals as diverse as mollusks and fish, individuals change sex during their lifetime. This is referred to as *sequential hermaphroditism*. Both female-to-male and male-to-female versions exist, with the choice thought to be due to the impact of body size on performance. For example, if females lay a huge clutch of eggs, they may benefit by developing later in life because they will be big enough. Conversely, in species where males are territorial and must intimidate others, they tend to develop from females so that they are large enough to compete. The reef-dwelling blue-head wrasse is a splendid example of the latter. If the single, large male is removed from the group of females he oversees, the largest female rapidly replaces him. In short order, she develops testes, turns a brilliant blue, and starts chasing the remaining females about. These diverse changes are coordinated by stress hormones, which spike early in the sex-change process in several species.[2]

How does a genetic or an environmental cue get converted into a male or female body? Biologists are shameless opportunists, and so historically we have used animals that are readily available and/or easy to rear to address this. As a result, the first organisms for which sex determination was understood at the molecular level were either lab-friendly and small (fruit flies and roundworms) or were ourselves (mammals, especially *Homo sapiens*). Each of these three pioneer organisms use a GSD system. For a long period, that was the only feature their sex determination circuits were known to have in common. The deep connections that were eventually revealed profoundly changed our view of animal biology.

THE FLY

Thomas Hunt Morgan was one of America's most illustrious biologists. He made his mark with his elegant studies of embryos and regeneration in the last years of the nineteenth century. Although he would eventually win the Nobel Prize in Physiology or Medicine for establishing modern genetics, Morgan entered the twentieth century as a leading skeptic of gene theory. At that time, genes were abstractions best understood from mutations affecting minor features of the organism, whereas Morgan and his embryologist colleagues could directly show that the dynamic matter in cells (the *cytoplasm*) had demonstrable control over development in different animals.

Morgan's conversion to a champion of genetics was led by the problem of sex determination and by an exceptionally talented doctoral student who joined his lab at Bryn Mawr College in 1900, Nettie Stevens. Stevens's dissertation on regeneration after injury impressed Morgan. Aware of recent research that suggested male and female animals may have different chromosome sets, Stevens next proposed a series of experiments to settle the matter. Morgan backed her, telling the Carnegie Institution of Washington that she had "an independent and original mind and does thoroughly whatever she undertakes."[3] It is worth noting, however, that in that same year, he had also dismissed the possibility of chromosomal sex determination in *Popular Science*.[4] Perhaps he expected Stevens's research to vindicate his position, but to his credit, he was nevertheless open to being proved wrong. In 1905, Stevens published definitive, microscopy-based proof that male and female mealworm beetles have distinct sex chromosome sets from the moment of fertilization of the egg by the sperm.[5] Stevens would die of breast cancer in 1912, just as she was offered the permanent research position she had so long sought. Nevertheless, her breakthrough would divert Morgan's research in a new direction, one that transformed the science of sex (and of so much more) forever.

In 1904, Morgan moved from Bryn Mawr to Columbia University, where he could focus exclusively on research. Not yet fully convinced that chromosomes could dictate sex, he studied several other insects, including aphids and related bugs.[6] This work revealed more cases of male and female chromosome differences. This question remained, however: were these the cause or a consequence of biological sex? Morgan continued to favor the latter. Around 1908, his lab began focusing on the fruit fly, *Drosophila*. Well-suited to lab life, flies would eventually offer a way to distinguish these two possibilities. By examining the results of countless crosses between flies bearing mutations he could track by eye, Morgan and his students extended Gregor Mendel's observations to account for the presence (and precise ordering) of many genes on a single chromosome.[7] In a landmark 1910 paper, Morgan showed that there are some mutations that have an effect when present in one copy in males but not when in one copy in females.[8] As more such sex-limited mutations accumulated, a plausible explanation was that males always have one X chromosome, and females always have two. Morgan was wavering.

To prove that sex chromosomes were the immediate cause of sexual fate, and not a consequence, an experiment that could create unusual sex chromosome combinations was needed. Another of Morgan's graduate students, Calvin Bridges, saw a way to do it. Using markers that could reveal extremely rare genetic events, he identified flies with two X chromosomes and a Y (XXY), and others with one X but no Y (XO). In 1916, he reported that XXY flies developed as females, while the XO flies developed as males.[9] This demonstrated that it was the number of X's, rather than the presence or absence of the Y, that makes the difference. Incidentally, Bridges's study was also the first time anyone had directly proven that chromosomes are the mediators of heredity—not too shabby for a rookie. Equally important for the field of developmental biology, it also showed that chromosomes can control a major developmental process.

The next obvious question was how a chromosome, which looks like a rather inert blob in the microscope, can actually *do* anything. It would be forty years more before the underlying structure of DNA and its iconic double helix was resolved, and seventy before the question would really be answered satisfactorily. However, an informative approach taken by early geneticists was to break genes at random and study the resulting mutations. No molecular-level understanding of what the gene did, or of what genes were more generally, was required to track mutations through crosses. Another member of Morgan's lab, Alfred Sturtevant, used this approach to provide the first indication that chromosomes contain specific sex-determining genes.

Sturtevant had developed a keen interest in heredity from pondering how coat color was distributed across the pedigrees of horses on his childhood farm. When he enrolled as an undergraduate at Columbia in 1908, it was inevitable that he would meet Morgan. He did, and he stayed on as a doctoral student. In 1913, he devised the method for using genetic crosses to place mutations in their order along a chromosome.[10] This was a stupendous use of indirect inference, but another of his findings is more important for our story here. As Sturtevant examined more mutations, he found one that seemed to blur the distinctions between male and female flies, which he named *intersex*. To his surprise, the *intersex* gene was not actually located on a sex chromosome. The process of determining sex was looking to be far more complex than had been imagined.

Sturtevant moved with Morgan in 1928 to help establish the first biology unit at the California Institute of Technology in Pasadena. As the world erupted into war in the late 1930s, many scientists saw their work put on hold. Having turned fifty-two weeks before Pearl Harbor was bombed, Sturtevant was protected by his age. This allowed him to continue his research as the fighting raged on, and shortly before the Normandy landings in 1944, an unusual bottle of flies presented itself. It contained many outwardly normal males that were infertile due to stunted testis growth. Follow-up crosses revealed that these sterile males lacked a Y chromosome and had the two X chromosomes normally found in females. Because of this near-perfect transformation of the XX fly into a male, he called the gene mutated *transformer* (abbreviated *tra*). As with *intersex*, *tra* was also not found on the X chromosome. Like other genes identified via mutations, they were named for the effect of breaking them. Thus, while XX *tra* mutants develop as males, the role of *tra* in a normal fly is to promote *female* development.

Work in other groups revealed two more key genes in the fly's sex determination system. One was reported in 1960 by the lab of Herman J. Muller, another alumnus of Morgan's original fly group. It was dubbed *Sex-lethal (Sxl)* because not only does its inactivation transform XX embryos into males (like *tra*), but it kills them for good measure. This added insult occurs because *Sxl* is the very first gene to interpret the presence of the XX or XY genotypes. Without *Sxl*, an XX embryo is unable to recognize that it has two X chromosomes. Acting as though it only has one, it hyperactivates all genes residing on the X (usually helpful in a male with only one), and this mistake is fatal. When genetic tricks are used to circumvent this lethality, XX *Sxl* mutants do indeed develop as males.[11]

The last crucial gene, *doublesex (dsx)*, was discovered in 1964 by Philip Hildreth in his University of California, Berkeley lab. Mutants lacking functional *doublesex* produced traits typical of both sexes at the same time, regardless of whether the mutant was XX or XY. This tells us that a key feature of sexual development is to block the features of males in a female, and vice versa.

The 1970s saw tremendous progress in molecular biology. In the 1980s, many labs around the world were applying these new tools to match long-known mutations with the actual genes that they altered. From this work, the genes noted above were finally identified at the molecular level. *Sxl* and *tra* encode proteins called splicing factors. Their job is to help the RNA copies of a gene that are produced when it is active, the messenger RNA (mRNA), to be processed properly. In eukaryotes (but not prokaryotes), the parts of mRNAs that contain the information encoding a protein (*exons*) are interrupted by stretches of RNA that don't encode anything (*introns*). Only when the introns are removed and the exons are spliced together can a functional protein be made. Sxl and Tra proteins affect the splicing of a small number of mRNAs, each of which are themselves involved in sexual development.

The gene *dsx* encodes a different kind of protein, one that binds DNA. In doing so, Dsx protein regulates whether genes with roles in female or male development are active or inactive, where "active" means it is transcribed into mRNA. Such gene-regulating proteins are referred to as *transcription factors*. Despite the distinct molecular functions of their protein products, a feature that the *Sxl, tra,* and *dsx* genes all share is that they produce mRNAs that are spliced differently in male and female flies. It turns out that this is the key to explaining how the fly turns a sex chromosome difference into a difference in anatomy.

Only female (XX) embryos have functional Sxl protein. This is a direct consequence of X chromosome number: the presence of two sends a signal that triggers the early production of *Sxl* mRNA. This early version of mRNA is different from those made later because it does not require special assistance to be spliced to produce functional Sxl protein. In contrast, embryos having only one X (like XY males) fail to make early Sxl protein. While both sexes go on to produce *Sxl* mRNA, that later version requires Sxl protein to be spliced productively. As a result, the early burst of functional Sxl protein found in female embryos maintains its own production by ensuring productive splicing of *Sxl* mRNA at later stages. Without the early Sxl protein, the mRNAs produced from the *Sxl* gene are all duds. Thus, only females can make Sxl long-term. This is an amazing example of autoregulation—when products of a gene regulate the activity of that same gene.

Sxl also regulates the splicing of *tra*, again promoting formation of a functional mRNA in XX females. Male *tra* mRNAs are thus also duds. A key mRNA target of the Tra splicing factor is *dsx*. In this case, however, both the XX version of *dsx* mRNA (produced in the presence of Tra) and the XY version (produced in its absence) have important jobs to do. The female form of Dsx (Dsx-F) inactivates male genes and allows female ones to be expressed. Conversely, the male form (Dsx-M) represses female genes but not the male ones. When the gene is completely inactivated, neither sex's genes are repressed, and the fly does its best to execute both the male and the female programs at the same time. This is how *doublesex* got its name.

The other mRNA whose splicing is tweaked by Tra is produced by the gene *fruitless (fru)*, which is only active in the fly's brain. Here, however, the logic of sex-specific splicing is reversed compared to *Sxl* and *tra*: in XX females their Tra protein ensures that all *fru* mRNAs are *nonfunctional*. In XY males, the lack of Tra allows it to be spliced productively. This generates Fru protein only in the brains of males, where it directs subtle changes to neural development and activity. Although XY males whose *fru* gene has been destroyed look overtly normal, they lose the ability to tell the sexes of their fellow flies apart. As a result, they spend considerable time courting other males, which explains how the gene got its name. This may seem to be a curiosity of flies, but as we will see in chapter 10, mutation of a gene involved in pheromone sensing (*TRPC2*) causes a similar behavior in mice.

OF MICE AND MEN

By the end of World War I, it was understood that girls and boys were distinguished by a genetic mechanism. Sex-linked genetic disorders, such as hemophilia, also indicated that females had two X chromosomes and males just one. Less clear was whether human males also have a Y, and what it meant for sex determination if they do. As with flies, evidence from both genetic crosses and microscopy was sought. But if fruit fly research was a deliberate and tidy intellectual exercise, then the effort to understand how human sex is determined progressed in an altogether different fashion. Consider the career of Theophilus Painter, whose transition from embryology to genetics paralleled that of Morgan.

In 1920, Painter was a young faculty member at the University of Texas at Austin (UT) looking for a way to make an impact after the disruption of World War I.[12] A former UT student, now a doctor at a nearby state mental hospital, would prove to be key to his success. One day, three men were brought with some trepidation to the hospital's infirmary. Each had been discovered masturbating by the staff and were to undergo "treatment" by the doctor. In dispatching his duties, the doctor understood that, to Painter, it presented an opportunity. It is not often that the methods section of a research paper is so suffused with human tragedy.[13]

> My material to date includes the testes of one white man and of two negroes. All individuals were castrated because of self-abuse, at one of the Texas state institutions. The testes were removed with the use of local anaesthetics and immediately preserved in Bouin's fluid, to which chromic acid and urea had been added. In less than a minute after removal from the body the germ-cells were being bathed in the fixing fluid. The preservation thus obtained is very satisfactory.[14]

However questionable the source of Painter's samples, they did indeed provide a clear indication that human males have one X and one Y chromosome. In addition, they enabled counts of the entire set of chromosomes. Painter noted that his clearest cases favored forty-six, yet he doubted his methods and suggested that forty-eight was probably correct. Forty-six was indeed the right answer, comprised of two copies of the twenty-two constant chromosomes (the *auto-somes*) plus the sex chromosomes (typically either XX or XY).

The same year Painter's study was published, apparent genetic evidence for the presence of a human Y appeared. A University of California medical student, Richard Schofield, published a short paper describing an odd pattern of inheritance of a funny trait in his own family, the webbing of two toes.[15] Schofield described it as always being passed down from father to son and never passed on to, or passed by, daughters. The eminent Harvard geneticist William Castle quickly recognized that it implied the trait was Y-linked and wrote his own paper (in *Science*, a big deal even then) pointing this out.[16] The existence of the human Y was now considered proven, and it seemed that Schofield had made an historic contribution to human genetics.

Good science requires skepticism, especially when a story seems too good to be true. The year after his paper was published, Schofield left genetics behind and started his medical career, but his story had become common knowledge. His family's toes were the first and tidiest example of a Y-linked trait, but Schofield was the only one who had observed them. In 1957, Curt Stern was a University of California, Berkeley researcher working across the bay from the San Francisco medical campus Schofield had attended decades earlier. Stern reviewed all purported cases of Y-linked human traits. Impressed with Schofield's evidence, he wrote to him to confirm the pedigree and asked to see some feet. Schofield was evasive and declined to arrange access to any of his family members.[17] His death in 1969 left the matter unresolved, but subsequent research on the genetics of webbed toes (which do indeed exist) suggests Schofield had misinterpreted his family's pedigree. Stern concluded that most traits put forth as Y-linked in humans had similarly shaky foundations. Today, the few human traits known to be caused by Y-chromosome abnormalities are mostly spontaneous mutations that cause immediate infertility so that they cannot be tracked through pedigrees.

With Stern's convincing dismissal of Y-linked traits, genetic evidence for a functional Y chromosome was once again sorely lacking. If no genes of importance are found on the Y, why does it exist? Does it even control sex? Fortunately for science, people are both very numerous and self-aware, and when afflicted with a genetic abnormality, they often walk into a clinic seeking help. In 1958, a young person of twenty-four who regarded himself to be male came to see Scottish geneticist Patricia Jacobs at Western General Hospital in Edinburgh. While he could pee standing up, so to speak, he also had noticeable breasts and atrophied testicles. The main source of testosterone in males is the testis (hence the name), so it is not surprising that he had a relatively high-pitched voice for a man and was unable to grow a beard.

Knowing that chromosomes were often involved in sex determination, Jacobs took a bone marrow sample from the man. By growing the cells in culture, she could examine the clearly separated chromosomes found only in dividing cells. She found that the man had a Y but also an extra X chromosome—he was XXY. This explained why he had some feminine attributes—the presence of a second X, normally found only in females, inhibits testis growth.[18] But it also implied something else: the early developmental decisions in the gonads and external genitalia to be male are determined by the presence of the Y and not the number of X's, as in fruit flies.[19]

A key, complementary observation that confirmed that of Jacobs was published the same year by a different group of British scientists.[20] They found that some women who fail to undergo normal puberty lack a second X and are thus XO. This also indicated that it was not the single X that initially produces a male but rather the presence of the Y. In females with only one X, however, later events of female development are abnormal, similar to the problems of XXY males. Also in 1959, American scientists at Oak Ridge National Laboratory showed that equivalent sex chromosome abnormalities had similar effects on the sexual development of mice.[21] That the Y was crucial for initiating male development across mammals was now clear.

What exactly makes the Y special? This would take great effort and another three decades to answer. One reason it was so hard is that the X and Y do not interact with each other in meiosis in the same way that the autosomes do. Autosomes pair up along their entire lengths during meiosis and exchange pieces prior to the completion of the two rounds of cell division (see chapter 1). This pairing requires that the two versions of the chromosome be largely identical. However, the Y has lost many genes and acquired large amounts of highly repetitive DNA sequences. As a result, pairing and exchange between the X and Y has largely ceased, only occurring at a small shared region at one end, the *pseudo-autosomal region*.

The method that Morgan and Sturtevant devised to order genes along a chromosome takes advantage of the fact that the farther apart any two positions are on a chromosomes, the more likely they are to be separated by an exchange event. If there were a single gene that determined maleness on the human Y (and, as you will see, there is), one would like to narrow down its location. However, if exchange between the X and Y is blocked, then all parts of the Y are invariably linked to the male-determining gene and thus provide no information. Without recombination mapping, a different, much harder approach would be required to identify the male-determining gene. Once again, it would depend on a supply of unusual people.

In every 25,000 babies judged at birth to be well-formed boys, one will actually have the female XX genotype. The first such XX males were reported by a Finnish team in 1964. Eighty percent of these have normal male genitals and harbor a portion of the Y chromosome attached to the X or to an autosome.[22] As Jacobs had observed from her XXY patients, males with two X's often seek medical help when puberty fails. More partial-Y XX males were identified, as well as XY females that lacked the part of the Y essential to trigger male development. These rare patients collectively offered a way to search for the minimal piece of the Y that was capable of specifying male development. The overall logic is simple: the smallest piece of the Y that is both necessary and sufficient for male development must be the key. Mice with similar chromosomal abnormalities provided additional clues based on the assumption (eventually proven to be valid) that the testis-determining factor (*Tdf*) is the same gene in most or all mammals.[23]

As molecular biology became more sophisticated in the 1980s and 1990s, researchers with different sets of informative patients raced to be the first to identify *Tdf*. One strong contender was a team led by the pioneering human geneticist David Page of the Massachusetts Institute of Technology (MIT). Page worked with Albert de la Chapelle, who had been part of the landmark 1964 XX male study, to determine which part of the Y had been carried over. He succeeded, and he also found that the same region, when deleted, produced XY females. There weren't many genes in the region, and a transcription factor gene in the region, called *ZFY*, seemed to be an excellent candidate. It was also found on the Y of other mammals. In 1987, Page's team tentatively proposed that *ZFY* could be the long-sought *Tdf*.[24] A media frenzy ensued.

Meanwhile, a London-based team headed by Peter Goodfellow and Robin Lovell-Badge were working with their own set of patients. Their work soon implicated a different gene very near *ZFY*, which they called *SRY*. Why did the teams reach different conclusions? Page's inference relied substantially on a single XY female, whose Y had a deletion that removed *ZFY*. But what Page had not initially realized is that this woman's Y had a second deletion that also took out

SRY, making it a candidate as well. Page published his discovery of this second mutation alongside the work of Goodfellow and Lovell-Badge in 1990.[25] Although cast by some media as the loser in a dramatic race, his forthright emphasis on the truth was exemplary and helped solidify the case for *SRY*. Page's lab went on to discover Y-linked genes required for male fertility and to assemble the first sequences of the highly repetitive mouse and human Y chromosomes, feats that are hugely important in their own right.

If *SRY* is the key to the Y, how does it make a male a male? *SRY* encodes a transcription factor protein (SRY) that turns other genes on by binding to their DNA. It is part of a family of related proteins that share a conserved motif, known as the "*SRY* box." Subsequent work showed that one of these, SOX9, is also a key testes-promoting factor. The *SOX9* gene is found on autosomes across vertebrates, even if they don't have a Y chromosome. *SRY*'s job appears to be to kick-start male development by activating the related *SOX9*. *SRY* arose early in mammalian history as a duplicate copy of *SOX9*, and its presence nudged the developing gonad toward male fate. The chromosome bearing this new copy became the Y. Initially it was a pretty normal autosome, but because of the lack of recombination with the X, the Y has been getting weirder and weirder ever since. Over time, as the Y lost functional genes, the number of X's also became a factor: male specification by the Y works very well when there is only one but not so well when there are two. Conversely, females with only a single X also have atypical sexual development.

THE WORM

By the mid-1960s, an international and interdisciplinary group of scientists had solved one of life's great puzzles—how the DNA of genes and chromosomes is read to produce proteins. Prominent among this group was Sydney Brenner, a scrappy South African from modest roots. Brenner's ambition and intelligence led to a quick rise through British science. He worked with Francis Crick to show that a triplet code converts gene language into protein sequence, and he worked in parallel with François Jacob and Matthew Meselson to demonstrate the existence of the messenger RNA intermediary that carries this code to ribosomes. With the mysteries of genetics fast clearing, Brenner began to set his sights on another inscrutable aspect of life: the brain.

By the end of World War II, the fruit fly was the king of animal genetics. But even the poppy seed–sized fly brain has over 100,000 individual neurons—a vast number for someone looking for a complete description. Brenner wondered whether a simpler animal might exist that could be as manipulable as *Drosophila melanogaster*. After a search and encouragement from a Californian nematologist named Ellsworth Dougherty, he settled on the roundworm (nematode) *Caenorhabditis elegans*. *C. elegans* adults have only about three hundred neurons, yet they are capable of complex behaviors related to feeding, movement, and mating. They also had an unusual sexual system, with typical males and female-like self-fertile hermaphrodites. Self-fertility made *C. elegans* much easier to work with genetically, while the optional males allowed the production of new gene combinations by crossing. Brenner was hooked.

French nematologist Victor Nigon had worked with *C. elegans* in the 1940s, at times with Dougherty. Under challenging conditions in wartime France, he demonstrated that hermaphrodite worms were XX and males XO in 1943. Beyond that, however, nothing was known about how the number of X's was converted into a male or female body when Brenner set up his worm lab in Cambridge, England. Brenner worked on several aspects of *C. elegans* genetics and development, but the problem of how sex was determined fell to his postdoctoral associate, Jonathan Hodgkin.

Hodgkin's family had more than its fair share of geniuses. His father and grandfather were both Nobel laureates; his mother, Lady Marni Hodgkin, was an important editor of children's books; and Hodgkin's lymphoma is named for a more distant relative. Hodgkin's choice to do his PhD on the little-known *C. elegans* with Brenner reflected his conviction that this tiny worm had the potential to reveal how behaviors emerge from the actions of cells and genes. He was drawn to the extreme differences in male and hermaphrodite behavior. As he put it, "adult males are sex obsessed and hermaphrodites are wholly uninterested; mating is a very one-sided business. Mating is also fairly complicated, perhaps the most elaborate part of the worm's behavioral box of tricks. So I decided to explore the basis of the difference between male and hermaphrodite."[26]

Hodgkin's first goal was to connect their behavior to their anatomy. He began by slicing the worms like minute salami and examining the anatomy of each slice with electron microscopy. What he found shocked him. "The male nervous system is decidedly more complicated than that of the hermaphrodite, and detailed reconstruction of the male neuroanatomy looked like a daunting prospect. Genetics was more fun. I began to study mutants with abnormal mating behavior or aberrant male development."[27]

When an XX hermaphrodite fertilizes her own eggs, all sperm bear an X and thus only more XX baby worms emerge. If a selfing hermaphrodite suddenly starts producing males, it's possible she has a problem with chromosome sorting in meiosis or that she carries a mutation that transforms the sex. When Brenner mentioned that he had frozen some mutants that seemed to be in the latter category, Hodgkin decided to investigate. As with flies, he expected that sex-reversed mutants can eventually reveal how each sex can develop from a common starting point in the fertilized egg. Four decades after Nigon's early studies, the work of Hodgkin and his associates would systematically reveal the circuit that responds to the number of X chromosomes. Along the way, they would also develop many useful tools for the study of *C. elegans*. Hodgkin also become the curator of gene names for the *C. elegans* community, a job that required both encyclopedic knowledge and the trust of his fellow researchers.[28]

When at last it was revealed at the molecular level, the worm sex circuit was nothing like that of the fly or the mouse. There were no mRNA splicing factors, no transcription factor related to SOX9. Instead, a protein on the surface of cells, TRA-2, conveys the female fate to a transcription factor protein acting in the nucleus, TRA-1. In XO embryos, however, a signal called HER-1 is produced, which binds to TRA-2 and inactivates it. As a result, TRA-1 is destroyed, and male development occurs. TRA-1 is a counterpart of *GLI*, a gene that is important in development in many other animals but does not function in sex determination outside nematodes.

While Hodgkin and his colleagues deciphered the signaling circuit that mediated sex, another group was working to understand how the sex chromosome count controlled the production of HER-1. This team was led by Barbara Meyer, a brilliant geneticist who, like Brenner, was looking for a new challenge. Meyer's lab discovered the gene that first responds to the number of X chromosomes, *xol-1*.[29] With elegant experiments, they showed that the X chromosome encodes repressors of *xol-1*, and the autosomes activators. One X doesn't generate enough repressors to overcome the activators, so *xol-1* is turned on in males. Once made, XOL-1 protein induces the HER-1 signal that conveys the male fate to surrounding cells. Two X's, however, can shut off *xol-1*, and we get no HER-1 (and hermaphrodite development) instead.

CONNECTING THE DOTS

A great discovery of 1980s biology was that a conserved set of genes controls the development of all animals. The poster children of this "common tool kit" were the *Hox* genes, which determine the head-to-tail pattern, and *Pax6*, which was central to the formation of eyes as different as those of humans, squid, and insects. By 1990, however, decades of careful research by various brilliant people had produced what seemed to be a clear, very surprising result: sex was different.

We can summarize the previous few pages as follows (figure 5.1): flies interpret sex chromosomes by a bucket brigade of regulated mRNA splicing, culminating in the production of sex-specific flavors of the Dsx transcription factor. Mammals use a Y-linked gene encoding a transcription factor, SRY, to directly tip gonad development toward testis by activating *SOX9*. *C. elegans* used a third approach, producing a protein signal in the XO embryo but not the XX, which determines whether a transcription factor of the Gli family, TRA-1, is present. These first three examples appeared to be wholly incongruous. Were male and female sexes invented independently in different phyla? If so, then our conclusion in the previous chapter (that separate sexes evolved only once in the Bilaterian animals) is wrong.

During this time of puzzlement, biologists marched onward into the fog. One of them was David Zarkower. Zarkower had joined Hodgkin's lab as a postdoctoral researcher in 1989 and led the efforts to isolate *tra-1*, the transcription factor gene described above. His success led to a faculty position at the University of Minnesota, where his first project was to locate the DNA sequence of a gene regulated by TRA-1, called *male abnormal 3 (mab-3)*. XO animals with mutations in *mab-3* were recognizably male, but the delicate tail structures they use for mating were messed up, and they produced egg yolk, something only hermaphrodites normally do. This was intriguing, but the incomplete effect suggested it could be a relatively minor player in sexual development. Zarkower recalls that when he shared his lab's efforts to clone *mab-3* with an eminent visiting researcher, he was advised to "ditch *mab-3* and find a better gene."

Zarkower declined to take his senior colleague's advice and kept pursuing *mab-3*. During this time, the *C. elegans* genome sequencing project (the first for any animal) was nearing completion, and this was making it far easier to connect a mutation with a gene—for most people, anyway. As Zarkower tells it, "We had

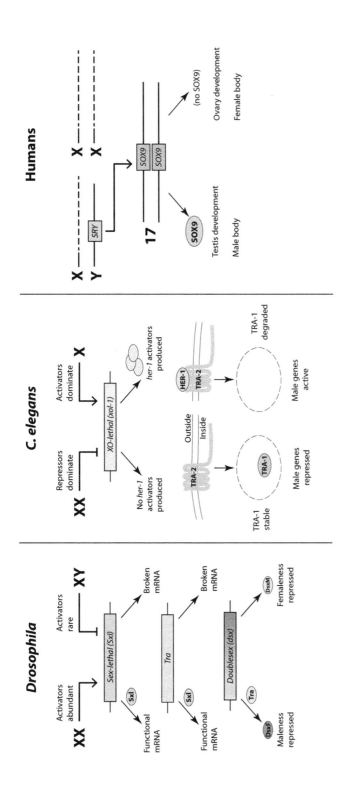

5.1 Animal sex determination pathways, circa 1990. Each diagram summarizes key cues and control points that mediate genetic sex determination, with many components omitted for simplicity. In *Drosophila* (*left*), the number of X chromosomes dictates how the mRNAs produced by the genes *Sex-lethal* (*Sxl*), *Transformer* (*Tra*), and *doublesex* (*dsx*) are spliced. As a result, sex-specific forms of Dsx transcription factor protein are produced. In *C. elegans* (*middle*), the expression of *XO-lethal-1* (*xol-1*) is controlled by the number of X chromosomes. In XO males, the HER-1 protein is produced, which leads to degradation of the TRA-1 transcription factor in cells that detect its presence via a surface receptor, TRA-2. TRA-1 prevents the male development that would otherwise occur by default. In humans (and most other mammals), the Y encodes a transcription factor, SRY. Its main function is to promote production of a related transcription factor, SOX9. Abundant SOX9 protein leads to testis formation and male development via gonad hormones.

a hell of a time cloning it, and had to resort to a trick: it was in one of the last gaps in the worm genome." With persistence the sequence of *mab-3* was finally revealed in 1998: it is the worm version of *doublesex*, and one of the key male genes that TRA-1 represses. I asked Zarkower to describe the moment he first knew:

> I was home working on the basement when Chris Raymond called. He just said "What would you say if I told you it was *doublesex*?" I said, "I'd say you'd better not be joking." I was so shocked I literally had to sit down. It was one of those moments where suddenly everything becomes clear—I immediately realized sex determination was a conserved process across hundreds of millions of years and that it was just as likely that a *dsx* homolog was regulating sex in mammals if it was there in worms. I felt like suddenly the horizon cleared and I could see a great distance and we knew something really important that we hadn't known literally seconds earlier. It was one of those extremely rare but thrilling moments we all work so hard for.

Zarkower's hunch that transcription factors of the Dsx-MAB-3 family (referred to as DM proteins) were widespread in animals was correct: sex finally had its own *Pax6* equivalent.[30] In humans, the SRY and SOX9 proteins act by turning on one of our several DM family proteins, DMRT1, in male gonads. In short order, DM proteins were identified in organisms as diverse as corals, sponges, and fish. In all cases, DM family members have been associated with sexual development, and most often (as in *C. elegans*) with male development.

Another clue that the DM proteins are deeply imbedded in sex determination comes from their presence in ESD species, such as turtles and sex-changing fish. Although there are many cues used to specify male or female development, they all exert their effect through DM proteins (figure 5.2). By 2000, the widely held impression of independent origins of sexes had given way to a different view: a primordial system of sex determination existed, but it has been radically modified in different animal groups, leaving only the DM family to reveal their deep historical connections. Why and exactly how sex determination systems change is still mysterious, but it is an active area of evolutionary research.

In chapter 4, we concluded that the first animals were hermaphrodites. These animals must have had DM proteins—what were they doing back then?[31] It is likely that they used DM proteins to distinguish their ovary from their testis, which are typically very different organs. In the male-female animals, this role

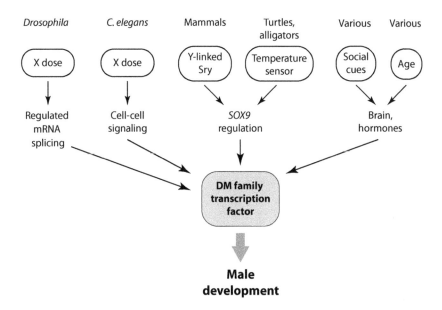

5.2 The doubleses-MAB-3 (DM) family reveals deep connections between different sex determination systems. Doublesex-MAB-3 (DM) transcription factors promote male development across animals in response to a wide variety of cues. *Drosophila* and other arthropods are exceptional in requiring the female-specific variant, Dsx-F, for female development.

in gonad formation was likely retained, and in many animals, the gonads are the main difference between sexes. Consistent with this, inactivation of *DMRT1* in only the gonad of XY mice transforms it into an ovary-like organ, while inappropriate activation of *DMRT1* in an XX gonad produces a testis-like transformation.[32] With the machinery for sex-specific gonad development in place, the stage was set for it to be connected (directly or via hormones) to other traits that differ between males and females, such as pigmentation, body size, and horns. In this way, the evolution of sperm and eggs, that great quid pro quo discussed in chapter 3, was eventually connected to the emergence of separate sexes.

CHAPTER 6

Going It Alone

Synopsis

Eggs provide most of the resources needed for an embryo. Partially or entirely cutting males out of the picture can achieve a big boost in reproduction, yet most animals continue to mate every generation. Sex persists because it helps species adapt to an ever-changing world in the short term, avoid mutational sickness in the medium term, and survive rare catastrophes in the long term.

IMMACULATE CONCEPTIONS

The first eight years of Anna the anaconda's life were chaste and generally fairly uneventful. She and her female roommates at Boston's New England Aquarium spent their days on display, occasionally getting a meal and resting a lot. Her keepers liked it that way, but she had a surprise for them. In the winter of 2019, she produced a litter of babies, three healthy and a dozen more that were stillborn.[1] It was clear from both Anna's closely monitored life and from subsequent genetic testing that no father was involved. That is, her progeny were the product of virgin birth (*parthenogenesis*), an immaculate serpentine conception that mashed up Eve's temptation and the Virgin Mary right there in front of everyone. The media went nuts, and men trembled as they contemplated such dispensability in our own species.

Anna's babies got attention because they are weird and rare for a snake, but for some animals, parthenogenesis is the norm. The aphids on your prize rose bush and the rotifers and water fleas (*Daphnia*) in your local golf course's water trap pond undergo multiple generations of parthenogenesis between rarer bouts of sexual reproduction, with the switch to mating triggered by the end of warm weather.

A somewhat creepier example of asexual reproduction is the parasitic nematode *Strongyloides papillosus*, which lives in the gut of sheep. Ensconced in innards, they are all parthenogenic females, presumably because it would be tough to find a male and mate in there. But reminiscent of *Daphnia*, these females can tweak their eggs to generate conventionally sexual male and female babies. These sexual offspring are plopped with a poop into the pasture, which provides a better place to mate. If there are genetically distinct females releasing sexual babies at the same time, novel genetic combinations can be produced. Even dung can be sexy, apparently.

The animals described above always reproduce via eggs, but they vary in whether the eggs are fertilized or not. There is another route to asexual reproduction that skips the egg entirely (figure 6.1). Many invertebrates reproduce between rounds of mating by budding off parts of themselves that can grow into full-size individuals. An example is the jellyfish, which pumps out many swimming medusae from a polyp anchored on the seafloor. Another is the acoel *Convulotriloba*. These simple worms routinely subdivide their bodies by developing new wormlets at their rear end. When large enough, they pop off to live their own lives. In some *Convulotriloba* species, the babies face backward relative to the big mamma worm, while in others, they face forward. As with the egg-dependent parthenogens, rarer sexual reproduction also occurs.

Some animals can only reproduce asexually. This can happen if they are hybrids between two species that have different numbers of chromosomes but otherwise the same genes. Such animals are healthy, but the mismatched chromosomes block the chromosome pairing process that is the first step of meiosis (see chapter 1). In others, an offspring with three sets of chromosomes is produced (an example of a *polyploid* organism, one with more than two chromosome sets), which again prevents meiosis from happening normally. In both cases, parthenogenesis is a spontaneous response that can keep a lineage alive, at least for a while.

For reasons given below, exclusive parthenogenesis is usually an evolutionary dead end. However, animals that use it can still be major players for a time, and presumably this is why they keep popping up. A case in point is the marbled crayfish. In 1995, a hobbyist caught an apparently normal slough crayfish (*Procambarus fallax*) in Florida and put it into a tank. It didn't look unusual, but it

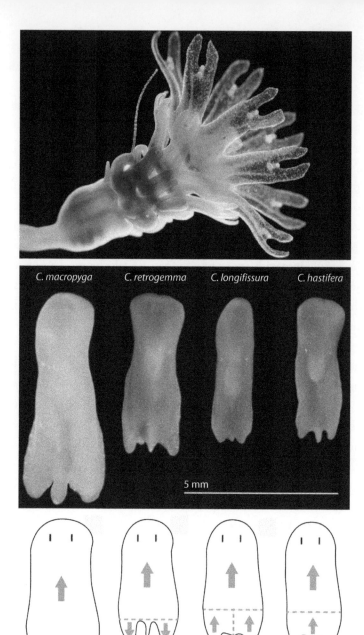

6.1 Asexual reproduction by budding. *Top*: A polyp of the jellyfish *Aurelia aurita*, preparing to release a series of flat juvenile jelly fish, called *ephyra*, into the water. *Middle and bottom*: Asexual budding in the acoel worm *Convulotriloba*. Four species are shown, with the body orientation of the reproductive buds at the rear indicated by the arrows. Budded offspring can be formed with the same or with opposite orientation as the parent, and they can form either one or two new worms.

Source: (*Top*) Ruben Duro, Science Photo Library; (*middle*) Thomas Shannon III; (*bottom*) redrawn from J. M. Sikes and A. E. Bely, "Radical Modification of the A-P Axis and the Evolution of Asexual Reproduction in *Convolutriloba* Acoels," *Evolution & Development* 10, no. 5 (September–October 2008): 619–631, https://doi.org/10.1111/j.1525-142X.2008.00276.x.

was actually a parthenogenetic female. Its asexual offspring quickly crowded the tank. The owner began offering them to pet shops and other crayfish aficionados, and soon the descendants of this unique female were everywhere. At some point, they were released into the waters of Europe and Madagascar and are now rapidly spreading. By genome sequencing, researchers in Germany found that the female and all her descendants are triploid.[2]

WHEN MOM IS ALSO DAD

In chapter 4, we saw another way for a single parent to produce offspring without a mate: self-fertilization. This is still sexual reproduction because both sperm and eggs are the products of meiosis that merge at fertilization. It is similar to parthenogenesis, however, because one individual suffices for reproduction. The evolution of selfing from strict mating is especially common in flowering plants, in which ancestral hermaphrodites shift from being self-incompatible to self-compatible by loss of a self-pollen recognition system. The tomato family (which includes potatoes, chili peppers, and other nightshades) alone has roughly one thousand examples.[3] Selfing is also seen in animals. As with plants, some were already hermaphroditic, and this presumably made it relatively easy to evolve.[4] There are also hundreds of examples of selfing appearing within groups that were ancestrally male-female.

The creation of a hermaphrodite from an ancestral female is a remarkable evolutionary innovation, both because of the developmental tweaks required to do it and because of its radical consequences. Sex within a single organism alters long-standing norms of inter- and intrasexual behavior as well as the way genetic variants move in a population. I study two examples of recently evolved selfing in my own lab at the University of Maryland, the nematode *Caenorhabditis elegans* and the mangrove killifish, *Kryptolebias marmoratus*. In both cases, former females evolved so that their historically all-ovarian gonads now produce sperm as well. These sperm can be used only for self-fertilization because the hermaphrodites lack the anatomy (in the case of *C. elegans*) and the behavior (for both species) required to provide them to another hermaphrodite. Males are also found in both, although they are much rarer than their ancestral 50 percent of the population.

Males in *C. elegans* form from embryos that have only one X chromosome. These are formed in crosses with an existing male or more rarely via selfing due to a meiotic error. *K. marmoratus*, however, makes males differently. Those of us caring for our killifish colony have all been surprised when a fish that had been brown and laying eggs one week was bright orange and no longer laying the next: they had changed sex in a brief time. That males develop from sex-changed hermaphrodites was first described in the 1960s by Robert Harrington, a pioneering biologist who spent most of his career working for the Florida State Board of Health in Vero Beach.[5]

Later experiments from the lab of John Avise confirmed that when paired with a hermaphrodite in either the field or the lab, they can fertilize the small portion of eggs that are not self-fertilized.[6] It is likely that *K. marmoratus* evolved self-fertility from a male-female ancestor with sex chromosomes (XX for female, XY for male). If so, the species may have spent time with no males at all because a population of XX hermaphrodites would have lost the putative Y chromosome. Exclusive selfing quickly approximates parthenogenesis—babies become genetically identical to their mothers. Consequences of asexuality described below may have selected for the ability to once again produce males without a Y. How this trick is achieved was studied by a doctoral student in my lab, Jax Ficklin. Ficklin found that, while administration of extra testosterone can turn hermaphrodites into fish that look like males, they never completely lost their ovary tissues, and upon removal of the hormone, they lose their orange color and resume laying self-fertilized eggs. This indicates that hormones are important, but something else is required to get the sex change to stick. What that might be is still a mystery.

WHY MARBLED CRAYFISH ARE EVERYWHERE

Some animals use parthenogenesis and selfing intermittently. They may self to produce progeny when mates are hard to find or when times are predictably good. The first situation is easy to understand: an organism must reproduce somehow if it is to continue as a species, and even crummy, inbred offspring are better than none. But what is it about benign conditions that promote asexuality or self-fertilization? To understand this, we need to do a thought experiment.

Let's compare an animal whose offspring are half males and half females, like the standard slough crayfish, to a variant that produces only parthenogenic females, like the parthenogenic female that spontaneously arose in Florida. The minimum reproductive set for each is either one male-female pair or a single parthenogenic female. We can take that and call it Generation One. If both types of female can produce ten eggs, the sexual parents will get together to make ten offspring, and the parthenogenic female also makes ten offspring. So far it seems even. But note that in the male-female case, five of the second generation's off-spring are males.

To produce the third generation, the five females each make ten babies, for a total of fifty. In contrast, all ten of the parthenogenic female's second generation offspring are females. If each of them makes ten babies, we have one hundred. This is a twofold difference—not subtle. And it's exponential, so that in the next generation the relative reduction in offspring number suffered by the sexual form is one half times one half, or one quarter the parthenogenic form (figure 6.2). By generation ten, the asexual form has produced over 250 times more baby crayfish, which means it should quickly take over. And that isn't even taking into account the ability of a single parthenogenic female to found a new population—a huge advantage for spreading into new areas.

Like parthenogenesis, selfing can also quickly spread through a population because sperm are cheap. If we assume the resources required to make just enough sperm to self-fertilize are much less than that needed to fertilize another individual reliably (either by mating between hermaphrodites or through males), then this should also be highly advantageous by the same logic. In lab cultures of *C. elegans* in which selfing is blocked by a mutation, the growth advantage of self-fertility allows rare second mutations that restore selfing to spread quickly.[7]

THE DOWNSIDES OF BEING A SINGLE PARENT I: THE SHORT-TERM

The existence of healthy one-parent reproduction in crayfish, *C. elegans*, and other animals poses a mystery: given the huge growth advantage it confers over strict mating, why didn't this system take over the planet a long time ago? You might think that the answer to such a fundamental question has been clear for a long time, but it has been surprisingly tough to test rigorously. The benefit(s) of sex must be both rather large and rather fast-acting. What might they be? Recent research suggests the dangers of reduced or nonexistent mating act on several different timescales.

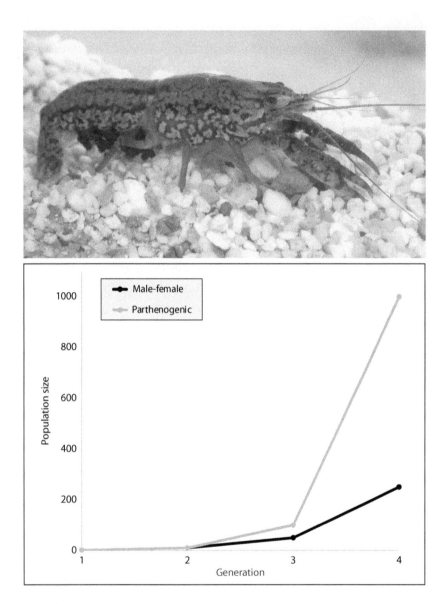

6.2 Crayfish demonstrate the short-term advantage of parthenogenesis. (*Top*) The parthenogenic marbled crayfish. (*Bottom*) Compared to its male-female progenitors, the parthenogenic variant is expected to enjoy an exponential growth advantage, assuming unlimited resources and identical egg production by sexual and asexual females.

Source: (*Top*) Steffen Harzsch (Universitat Greifswald).

The most immediate threat to selfing lineages is the phenomenon of inbreeding. As discussed in chapter 4, self-fertilization allows damaged (mutated) copies of genes to comprise both copies (be *homozygous*). The resulting sickness and death are two reasons outcrossing is hard to shake. In contrast, truly parthenogenic lineages preserve both sets of chromosomes from the founder, with no increase in homozygosity. Thus, inbreeding depression for parthenogenesis is not a factor, and they get a free pass (for a while).

With the right mix of selfing, outcrossing, and luck, a number of self-fertile organisms have managed to become entirely homozygous for all genes and still remain quite healthy. Mustard family plants like *Arabidopsis thaliana*, nematodes like *C. elegans*, fairy shrimp, and *K. marmoratus* have all managed this feat. Harrington demonstrated the homozygosity of *K. marmoratus* in a particularly clever experiment: fish of a given lineage can serve as organ donors for each other without any immune rejection.[8]

The above suggests a selfing variant that purges its nasty mutations is now justified in performing an elaborate end-zone dance.[9] Alas, that would be premature. Mating hugely accelerates the production of new combinations of gene variants, and this ability can be essential to counter the ever-changing threats that the organism faces. This has been documented in organisms as diverse as yeast, snails, and flies, but I think the most elegant demonstration was carried out by my colleague at the University of Oregon, Patrick Phillips, and two former graduate students in his lab, Levi Morran and Michelle Parmenter. They took advantage of some unique attributes of the nematode *C. elegans*. Populations of these tiny, fast-growing animals are usually a mix of abundant self-fertile hermaphrodites and rare males. Because hermaphrodites cannot mate with each other, males represent the only way genes can be exchanged between individuals.

Over decades of research, geneticists have developed mutant strains of *C. elegans* that affect reproduction. Some cannot produce any males (because they die as embryos), and in others the hermaphrodites are spermless and are thus true females that must mate with males to reproduce.[10] The degree of outcrossing found in *C. elegans* thus ranges from zero (in the hermaphrodite-only situation) to constant (in the case of the female-male mutant), with normal worms being intermediate. The Phillips group used this variation to see how outcrossing level affected the worms' ability to contend with two challenges: chemicals that damage their DNA and a lethal bacterium called *Serratia*.

When grown with the DNA-damaging chemical or the pathogenic bug, the strain that couldn't produce males (and thus couldn't produce new genetic combinations) quickly succumbed to both, but the two sex-capable strains did not. Even more interesting, in the hermaphrodite-male strain, males went from about 1 percent of the population to nearly half. That is, when times got tough, the frequency of males became similar to that seen in the male-female species that are *C. elegans*'s closest relatives. This indicates that, if things get bad enough, the crossed offspring sired by males were so much healthier that they made up for the reduction in egg production that would keep males at a low frequency in good times. This landmark study showed how outcrossing can be hugely beneficial on short timescales.[11]

In a follow-up study that Morran conducted with Indiana University's Curtis Lively, he found that when the pathogen is allowed to coevolve with the worm, males remain abundant indefinitely.[12] Thus, we can conclude that environments that are consistently challenging will select against low- or nonmating lineages, favoring those sexual forms that can adapt most rapidly. At some point, a strictly asexual organism like the marbled crayfish will have a real problem on its claws, and that will be the end of it.

THE DOWNSIDES OF BEING A SINGLE PARENT II: MULLER'S RATCHET

It is conceivable that once a selfing species has conquered all available habitats, it is so widespread that at least some local populations will never encounter a threat that requires rapid adaptation. *Now* can they do the end-zone dance? Nope. Even an initially healthy homozygote or parthenogen will gradually accumulate new mutations through replication errors and DNA damage. There will be many of these mutations eventually, and without meiosis and mating, it can only get worse. This phenomenon is referred to as Muller's ratchet, after the American geneticist and Nobel laureate Hermann Muller.

How mutations inexorably accumulate deserves a bit of explanation. If an asexual species has an essentially infinite population size, then there will always be a few individuals that lack new mutations of consequence. This pristine founding genotype could be maintained by sheer numbers indefinitely. But this assumption of infinite population size is not valid, even for mosquitoes. There is a finite number of unmutated individuals, and their number will fluctuate every generation because of the vagaries of life in nature. In addition, the impact of a new mutation may be so subtle that natural selection is not able to pick individuals that carry it out of the crowd. At some point, the totally unmutated genotype will blink out, and once it does, it cannot return. This irreversible loss of the least-mutated genotype represents a click of the ratchet.

While the first few clicks of Muller's ratchet may go unnoticed, eventually even the least-mutated genotype will begin to show signs of sickness. The root cause of this is the fact that, when sprinkled randomly across a genome, mutations are much more likely to break something than to improve things. While a beneficial mutation may indeed arise and even spread, it cannot stop the ongoing acquisition of others that are harmful. The action of Muller's ratchet means that, while strictly parthenogenic creatures like the marbled crayfish are enjoying their moment now, eventually they will succumb.

We can now see how complete commitment to selfing or parthenogenesis is a very different proposition than a mixed system in which one-parent reproduction alternates with mating. Even rare sex is enough to counter Muller's ratchet. Consistent with this, many parthenogenic organisms, like aphids, insert an occasional sexual cycle. Outcrossing is also retained in selfing animals. For *C. elegans* and *K. marmoratus*, whose hermaphrodites cannot mate with each other, this is mediated by rare males. These mixed strategies couple the growth advantages of male-free reproduction with the therapeutic benefits of outcrossing.

THE DOWNSIDES OF BEING A SINGLE PARENT III: THE LONG RUN

Not all species face daily crises like those to which the worms were subjected in the above experiments. The dandelions in your lawn are parthenogenic, and they thrive in that simple ecosystem. With a stable environment and a winning gene combination, one-parent reproduction could last for a long time. So long, perhaps, that sexual relatives would be pushed to extinction before the benefits of sex can kick in. We don't see this, however—why?

One important clue has come from the tomato family of plants. In 2010, University of Illinois, Chicago, biologists Emma Goldberg and Boris Igić and their collaborators examined the impacts of self-fertility on longer-term evolution.[13]

Using sophisticated statistical methods, they found that once selfing evolved in a lineage, it was less likely to produce new species and/or more likely to go extinct. A similar dynamic appears to occur in some groups of nematodes.[14]

The tomato study makes in important point: Even if a species can find a way to produce healthy offspring without constant mating, it is usually a Faustian bargain: that which is great in the short and medium terms can kill in the longer term. While eschewing mating can work for thousands or millions of years in some species, the *C. elegans* experiment above shows how an environmental change can require an organism to adapt quickly via novel gene combinations. Sex is very good at producing these from the set of existing variations, but asexuality can respond only if new mutations happen to come along. This much slower response presumably can doom a completely asexual lineage.

A species that doesn't mate could, in principle, be saved by re-evolving outcrossing again. In practice, however, this is very hard. Self-fertile mustard plants got that way because they have function-killing mutations in the entire gene set required to enforce outcrossing. This would be very hard to reverse. Work carried out in my own lab (discussed further below) found that a self-fertile nematode quickly shed thousands of genes after it switched mode, including some essential for effective mating.[15] It would thus appear that species that adopt selfing only have a limited time to turn back before the door to regular outcrossing is closed forever.

BDELLOID ROTIFERS: A SCANDALOUS EXCEPTION?

Between inbreeding depression, slow responses to new challenges, and Muller's ratchet, all exclusively nonmating species should eventually be wiped out, and it appears that they usually are. However, a few apparently long-lived asexual species have been described. The most famous are a group of rotifers, microscopic animals that live in ponds; puddles; and even tiny, temporary drops of water. DNA sequence differences and fossil rotifers found in amber suggest they have lived without any mating between individuals for at least 25 million years. Is this possible? If so, then how do they do it?

When researchers at Harvard and the Marine Biology Laboratory in Woods Hole, Massachusetts, scrutinized the genomes of bdelloid rotifers, they found a number of odd features.[16] First, many of the genes located at the ends of chromosomes were recently derived from the DNA of nonrotifers, such as bacteria and fungi. Some are likely functional because they were transcribed into mRNA and don't harbor obvious inactivating mutations. Some of the bacterial genes had also acquired features (like introns) that make them structurally more like those of eukaryotes. This indicates that bdelloids have a tendency to pick up DNA from their environment and, in some cases, use it.[17]

Second, bdelloids often have four copies of each gene in their genome.[18] This should reduce the speed of Muller's ratchet because it will take longer to break all of them by random mutations. It has been proposed that exchange between these copies (without mating) may achieve many of the same benefits of sex, especially with regard to evading Muller's ratchet. Is this kind of "parasexuality" why bdelloids are still around? As the methods for sequencing and analyzing genomes have become more sophisticated, it has become increasingly doubtful that bdelloids are truly asexual.

One cause for suspicion is that they retain the complete repertoire of genes required for meiosis. In 2020, more direct evidence for recent meiotic recombination was published by a group of Russian and American geneticists.[19] Olga Vakhrusheva and her colleagues found that genetic variants at different sites were not locked into long-term static combinations as one would expect if there were no sex. Instead, a variant at one site was increasingly likely to become separated from a variant at another site as the distance between them increased. This is a classic expectation of standard meiotic sex. Two years later, the lab of Harvard's Matthew Meselson, an early proponent of the asexuality hypothesis, replicated Vakhrusheva's result in a second bdelloid species.[20] As of this writing, it appears that some or all bdelloids may engage in rare sex. That mating has thus far not been observed is still hard to explain—perhaps these tiny animals really like their privacy.

THE GENOMIC BASIS OF MOJO

In chapter 5, we saw how genetic circuits specify male or female bodies in species with separate sexes. Despite their variety, eventually they all funnel down to transcription factors, key regulators of gene activity. Transcription factors exert an impact on a cell or organism by determining whether other genes' DNA is copied (or "transcribed") into a messenger RNA (mRNA). The abundance of a gene's mRNA transcript is thus one measure of its activity. When genome-wide comparisons of mRNA expression became possible in the late 1990s, many genes were found to be expressed to a greater extent in one sex than the other. For example, the lab of my National Institutes of Health (NIH) colleague Brian Oliver found that about 7 percent of fruit fly genes are at least twice as active in males, and about 4 percent at least twice as active in females.[21]

Work on *C. elegans* from Yale's Valerie Reinke found that these nematodes were similar to flies in their fraction of sex-biased genes, but they did not share the excess of male-biased ones.[22] The apparent difference between the fly and worm was notable, but a few years later it essentially evaporated. A grad student in my lab, Cristel Thomas, worked with Oliver's lab to repeat the *C. elegans* sex comparison. This was part of a study examining a larger set of related worm species, and it used more sensitive RNA sequencing methods.[23] Cristel's work revealed that male-biased genes are roughly twice as common as female-biased genes, as they are for flies. In addition, advances in technology allowed us to say with confidence that some genes have activity that is absolutely sex-specific, and again these are more numerous in males. The general finding that males have more sex-biased and sex-specific genes seems to be general across animals with separate sexes, including humans.[24]

Of all those sex-biased genes, what fraction are dedicated to the basics of making and delivering eggs and sperm, and what fraction to related tasks, like finding a mate or fending off competitors? Selfing species that have recently spun off from male-female ancestors can allow us to answer this question. While they retain the core gamete production machinery, their attractiveness to prospective mates is diminished, and their mating behavior becomes less reliable.[25] The genetic basis of this erosion of sexual prowess was examined by another grad student in my lab, Da Yin, in collaboration with Cornell University's Erich Schwarz.[26] Yin and Schwarz compared the genomes of another recently self-fertile *Caenorhabditis* species, *C. briggsae*, with that of its male-female sister species, *C. nigoni*. They found that *thousands* of genes present in the last common ancestor of the two species were lost in *C. briggsae*. Consistent with the change in sexual mode in the *C. briggsae* lineage, the majority of these encoded proteins produced mostly or only by males.

In a species in which mating is optional, perhaps it is not surprising that sex-related genes would atrophy. However, what Charles Darwin called "loss through disuse" is not sufficient to explain the rapid changes we observed. Instead, it seems that some genes that historically promoted male fertility are actually pushed out of the genome by selection once selfing evolves.[27] Freed from inbreeding depression by purging (see chapter 4), selfing nematodes can actually reproduce faster if males are not very good at siring offspring. This is because virgin XX hermaphrodites beget only more hermaphrodites, whereas half of the offspring fathered by an XO male will be male because of the presence of sperm without an X chromosome. Thus, in benign conditions, it behooves the hermaphrodite to eschew mating, like marbled crayfish. However, because natural populations of *C. briggsae* and *C. elegans* still have rare fertile males, preserving the ability to mate appears to be important for the long-term survival of the species.

SUMMING UP

In this chapter, we have seen that when eggs provide most of the resources for a new organism, mating every generation can greatly slow reproduction relative to asexuality or self-fertility. Many animals use one-parent reproduction much of the time, but they switch to mating when times get hard. Achieving this by the path of self-fertilization is slowed by inbreeding depression, yet many lineages have eventually gotten around this through purging. One-parent reproduction by parthenogenesis or budding does not incur inbreeding depression, and animals that do this some or even all of the time have become extremely successful. However, species that stop mating altogether will eventually go extinct, as predicted by Darwin in 1859, in the fourth chapter of *On the Origin of Species by Means of Natural Selection*.

In the century and a half since Darwin wrote, evidence has accumulated that sex is promoted in the short run by the need to adapt to an ever-changing world, in the medium term by the ability of sex to prevent the accumulation of harmful mutations (Muller's ratchet), and on the scale of millions of years by singular events that disproportionately wipe out species that have adopted reduced sexuality. Because the penalty for adopting an asexual strategy may only be enforced long after it evolves, this creates a dynamic in which species that convert to asexuality will appear, prosper obliviously for a time, and then die out. Although some of this chapter's content (and some sensationalist claims by those who should know better)[28] might at first suggest the days of mammalian males are numbered, I conclude we have a solid (if potentially rocky) future.

CHAPTER 7

Land Ho!

Synopsis

Our fish ancestors released their eggs and sperm into the water. As some moved onto land and became amphibians, they continued this tradition. Amniotes overcame the dependence on water with clever inventions (like penises and eggshells) that kept sperm, eggs, and embryos wet. This reproductive superpower allowed them to spread rapidly around the planet.

———◆———

My sneakers crunched rhythmically on the gravel as I made my way up the dry wash. I was a few miles from the lower Colorado River, enjoying a late spring hike. The sky was its usual spotless blue, and the infrequent rain in the Mojave Desert makes for the kind of landscape that geologists dream of: mile after mile of mountains, canyons, and valleys with few pesky plants to cover their beautiful brown, purple, beige, and maroon colors. The land's tortured history of sedimentation, uplift, erosion, and volcanic activity is all right there, defying you to make sense of it. There is still plenty of life in this desiccated landscape, however. Along the rocky banks, a few wooly daisies were opening their bold yellow flowers, and wiry creosote bushes were sprinkled across the surrounding landscape. Among vertebrate animals, lizards, snakes, birds, rodents, and tortoises are especially abundant. Amphibians, not so much.

It had rained a few times since the new year, but in a few weeks, the last damp spots along the shady side of the channel would vanish in smothering heat. As

I rounded a bend, what was left of a puddle a few feet across came into view. I bent down to discover a sad tableau of mass tadpole mortality. One hundred or so whip-tailed, jet-black blobs were cradled in the drying mud, dead or dying and on their way to becoming pollywog jerky. They were optimistically produced by a pair of red-spotted toads a couple of weeks before, and a few even had managed to start sprouting their hind legs before the dry warm air, breezes, and lack of more rain doomed their bid.

The majority of desert toad broods end up this way. Still, they persist because the adults wait out the dry spells in cool rock crevices and can live into their teens. Some years in the American Southwest are unusually wet, and even one successful brood in a lifetime is enough to replace the parents. Like those of fish, the eggs of frogs and toads lack a waterproof shell to protect the developing embryo. And once hatched, the tadpoles spend their first few weeks of life in water, slowly growing and remodeling their bodies for life on land. This double requirement for water greatly limits where they can live. It's tough to be an amphibian in the desert, but even in much wetter climates, they must choose carefully.

Don't get me wrong—ancient amphibians definitely merit being honored in some future evolution hall of fame for being the first vertebrates to colonize land. But as my desert hike demonstrated, the requirement for standing water for the first few weeks of life makes many habitats marginal or totally off-limits for their descendants. In contrast, rattlesnakes, tortoises, kangaroo rats, and roadrunners have no such limitations on where they can reproduce. Practically every creosote bush has a resident lizard, even on a rocky hilltop. All these animals are part of a larger group called the amniotes, a group that includes us.

Amniotes are descended from amphibian ancestors, but over time they evolved a sort of reproductive superpower: the ability to mate and rear young away from water. The first amniotes surmounted three key challenges. One was getting the sperm and egg together without releasing them outside the body. Another was keeping the embryo wet until it is ready to hatch. Last, aquatic larval stages, like tadpoles, were eliminated so that the entirety of life was lived on dry land. By solving each of these puzzles, amniotes spread rapidly to even the driest land areas. From rocky cliffs to sand dunes, from jungles to savannahs, to Antarctic ice, they invaded the entire planet. How?

STEP ONE: INTERNAL FERTILIZATION

Although urinary freedom might seem like reason enough, it's finally time men learned why they have that thing flopping between their legs. In short, evolution has gone to great lengths to meet the stiff requirement that vertebrate sperm be kept wet and happy until they come to the egg. When you live on land and far from water, penises and vaginas are really good for that.

The first amniote fossils date from the late Carboniferous Period, roughly 320 million years ago. One might think that if the main job of the penis is to get the sperm and eggs together, then penises should all have remained more or less the same since then. But if we look across the amniotes (figure 7.1), we see a surprising variety of forms. For example, snakes and lizards have a forked penis composed of two mirror-image *hemipenes*. Impressive as this double-barreled member is, male snakes use only one at a time—the one on whichever side happens to be

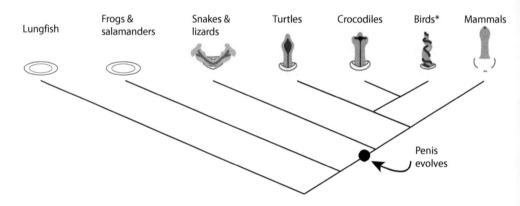

7.1 Origins of the penis in amniotes. Our closest fish relatives, the lungfish, and amphibians use a simple genital opening to release sperm into the water. As they moved away from life around water, the first amniotes (arrow) evolved a penis, and most of their living descendants have them. Birds get an asterisk (*) because 97 percent of them no longer have them. The black area in each cartoon represents the sulcus, or channel through which sperm travel. Only in mammals is this fully closed. The two embryonic components of the penis remain separate in adulthood in snakes and lizards, but are united in other groups.

Source: Modified from M. L. Gredler, C. E. Larkins, F. Leal, A. K. Lewis, A. M. Herrera, C. L. Perriton, T. J. Sanger, and M. J. Cohn, "Evolution of External Genitalia: Insights from Reptilian Development," *Sexual Development* 8, no. 5 (2014): 311–326, https://doi.org/10.1159/000365771.

facing the female.[1] In turtles, crocodiles, and mammals, the penis develops into a single organ. Single or double, however, they all develop from a similar embryonic starting point. In early development there are two patches of tissue that give rise to the penis. In snakes and lizards, these remain distinct into adulthood, but they fuse in the other groups.[2] We encountered the result of this fusion, the genital tubercle, in human fetuses in chapter 2.

Even within the familiar non-split-penis format, there is tremendous variety. For example, in most mammals, the penis is stored internally in a deflated state and is everted only for mating. For some reason, we humans have our flaccid penises constantly exposed to the slings and arrows of outrageous fortune. If we needed any more evidence that we evolved rather than having been created optimally by an omniscient benefactor, there you have it. Humans are also unique among the Old World monkeys and apes (our closest relatives) in lacking a penis bone, the baculum.[3]

Researchers led by Stanford University's David Kingsley have investigated another recent change in our species' penis. Chimpanzees and more distantly related mammals have sensory spines at the base. You probably are aware (and perhaps glad) that this is not the case in humans. Men (at least, *some* men . . .) have great stamina in intercourse relative to other great apes, and the decreased sensation from the loss of the spines may be a big reason why.[4] Kingsley's group found that our species' spineless state exists because of a mutation in the gene that encodes the receptor for testosterone, the male-promoting hormone. This regulatory mutation doesn't affect the receptor protein per se, but it prevents penile tissue from producing it at all late in the development of the organ.[5] Even though testosterone is abundant, it can no longer stimulate spine development.

Even stranger is that most birds have lost their penis altogether. They still manage to retain internal fertilization by a practice known as a cloacal kiss. Both males and females can evert the lower part of their cloacas. In doing so, they turn the inner side out, like a glove peeled off a hand. This exposes sperm on the surface in the male, and the inner reproductive tract in the female. When they touch, and then revert to their normal internal position, some sperm are transferred to the interior. The fact that birds lacking penises descend from ancestors that had them can be seen in their embryos,[6] where the paired rudiments that give rise to them form and begin to develop but then regress because cells are induced to die. When University of Florida researchers led by Martin Cohn blocked this signal in male chick embryos, they partially regained their penis.

Ducks are one of the few birds that do retain a penis, and they use them in a rather extreme way. They are unusually large for the size of the drake and have an unmistakable helical twist to them, like a corkscrew. Curiously, while a female duck's vagina is also helical, it has the *opposite* handedness. That is, female ducks do the opposite of helping. Researchers from Yale have concluded that this mismatch between parts evolved to protect females from the harmful effects of mating.[7] Using high-speed video recordings, they found that "eversion of the 20 cm muscovy duck penis is explosive, taking an average of 0.36 s." This force, if left unopposed by the female reproductive tract, would really hurt.

That there should be any of this sort of risk in mating may seem odd and counterproductive. Don't mates need to cooperate to be parents? They do, but each sex has somewhat conflicting demands upon it with regard to reproduction, especially when stable mate pairs are not formed and females do the rearing of offspring. As Charles Darwin first realized, males that are attractive to females and/or effective competitors against other males will come to dominate populations (see chapter 11 for more on this). This sexual form of selection has been found to explain many male traits, including ornaments and associated courtship behaviors—the stuff leading up to mating. But as Geoff Parker first noted over fifty years ago, competition doesn't always end there. If females mate with multiple males in quick succession, then their reproductive tract becomes part of the field on which their mates compete as well.[8] As a result, genitalia, testis size, sperm number, and even the motility of individual sperm are also subject to sexual selection. Thus, males are not *trying* to hurt their mates, but post-mating competition can lead to harm anyway. Females need fathers for their eggs, but they also need to preserve their ability to rear babies and produce future broods. Countermeasures that limit this harm to females thus evolve in response. How this sexual conflict shapes the evolution of genitals could easily be (and has been) the topic for an entire book.[9]

So bird penises were lost during development in many species, but when they are retained, there is often extreme sexual selection. So *why* did most birds lose them altogether? One possibility is being light is important for flight, and willies have weight. However, ducks and geese undertake some of the most arduous migrations known every year, and they have especially big ones. An alternative explanation is suggested by the strong sexual conflict found in birds that retain penises. If female birds evolve sufficiently effective ways to resist forced copulation, eventually the penis becomes a large, heavy organ that still doesn't guarantee male success. In this situation, behavioral changes that restore a more consensual form of mating would benefit both the male and the female. With these in place, the penis becomes unnecessary and soon disappears because the resources thus freed can be better used elsewhere. Biologist Patricia Gowaty has also reviewed evidence suggesting that forced mating produces offspring that are less healthy than those produced by consensual mating.[10] Why this should be the case is unclear, but if true, it would also select for cooperative mating and the loss of the penis.

STEP TWO: THE SHELL GAME

Salamanders, frogs, and toads are abundant in moist places, but they require extreme adaptations to survive in drier habitats. You won't find any salamanders in the Mojave Desert and, as we saw, the toads that live there spend most of their lives in hiding, waiting for rain or cooler temperatures. To gain access to the whole planet, a more universal solution is required to keep embryos wet. This solution was the eggshell and the membranes that package the amniote embryo within it. As you may have noticed while cooking breakfast this morning, the eggs of chickens (and other amniotes) have a mineralized shell (figure 7.2). It is made of crystalline calcium carbonate, with some proteins that bind it together.[11] The shell has the amazing property of allowing gas exchange through thousands of tiny pores while still preventing germs from getting inside and conserving the precious water within.

That water does get gradually lost became apparent to me when I found a long-forgotten dozen eggs in our basement fridge. Given the year of neglect that had elapsed, I cracked one open with some trepidation. I was not greeted by a horrible stench—the antimicrobial shell protected the contents from bacteria. But I did notice it was oddly light as I picked it up, and upon cracking it, I could see that it had lost about two-thirds of its volume. In a fertilized egg, oxygen and carbon dioxide need to move in and out (respectively) of the shell to meet the needs of the developing chick. My shrunken eggs showed that water molecules also move through the intact shell. Fortunately for chicks, this occurs slowly enough that they are normally hatched before it becomes a problem.

Flexible and tough protein membranes that line the inside of the shell also help the egg do its jobs. The outermost layer of the eggshell is the cuticle, formed

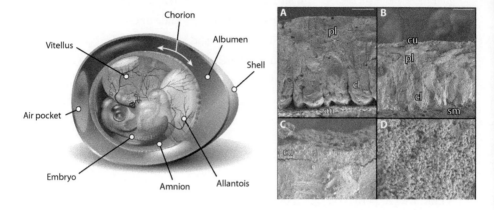

7.2 The amniote egg, an escape pod from the pond. Unlike in amphibians, egg-laying amniote embryos develop inside a protective shell that keeps them wet and germ-free (*left*). On the right, we see cross-sections of the eggshell of the chicken (*A*) and the Australian bush turkey (*B*), which nests in bacteria-rich compost. The bush turkey has an unusually thick cuticle (cu), composed of many calcium carbonate nanoparticles (progressively magnified in *C*, *D*).

Source: (*Left*) KDS4444, Wikimedia Commons; (*right*) panels from L. D'Alba, D. N. Jones, H. T. Badawy, C. M. Eliason, and M. D. Shawkey, "Antimicrobial Properties of a Nanostructured Eggshell from a Compost-Nesting Bird," *Journal of Experimental Biology* 217, pt 7 (April 2014): 1116–1121, https://doi.org/10.1242/jeb.098343.

from millions of tiny mineralized nanospheres that are stuck together. In habitats that are relatively clean, it is often lost, which tells us its main job may be to keep germs out.[12]

There is a curious disconnect in the fossil record between when the first amniote skeletons are found and when fossilized eggs appear. The former show up over 300 million years ago, but the later do not until about 190 million years ago. Does this mean amniotes weren't really laying eggs for a third of their existence? Maybe, but an international team led by paleontologist Robert Reisz recently offered an alternative. The team examined the very oldest dinosaur eggs known and found that they are much thinner than those at their peak in the late Cretaceous.[13] This trend toward thickness, when extrapolated backward, suggests that the first amniotes had even thinner shells, perhaps too thin to be preserved as fossils. Such thin-shelled eggs are still produced today by lizards, snakes, and turtles.

In chapter 1, we saw how photosynthesis first oxygenated the atmosphere, triggering the evolution of eukaryotes and sex over 1 billion years ago. But if you look again at figure 1.1 from that chapter, you will see the levels fluctuated greatly thereafter. After a huge spike in oxygen around the time of the Cambrian Explosion, it fell steadily until about 200 million years ago. At that point, it gradually increased again, a period in which the dinosaurs are diversifying and getting bigger. As eggs get bigger, they have less surface area to service each gram of embryo inside (see chapter 3). Even worse, the shells need to be thicker to support the weight. Reisz proposed that the thickening seen in the eggshells of later dinosaurs was possible only because atmospheric oxygen levels rose. More oxygen could have compensated for the reduced gas exchange of a thicker shell. This is certainly quite plausible, although gas exchange will also be affected by nest structure, temperature, and the metabolic rate of the embryo.

Eggshells have obvious advantages, but they also present logistical problems. For example, how do you put a hard shell around an egg and still fertilize it? The familiar chicken hen has been the model for biologists here and, as you might guess, the order of events is key.[14] In her ovary, she produces the huge yellow egg cell we call the yolk. If the hen has mated, the yolk is fertilized soon after it tumbles into the oviduct by sperm that have made their way up the female reproductive tract. Because the yolk is so large, the first cell "divisions" don't actually divide it completely. Instead, cell division is pushed to one side, so that the result is a disc of small cells along the surface. As the embryo forms, it is carried ever closer to the exit. Along the way, it is swaddled in albumen (or "egg white") and then given two protein membranes around that. Thus enclosed, the still-flexible egg enters an extraordinary organ called the shell gland. There, it is bathed in a saturated calcium solution that rapidly crystallizes into a rigid shell, and then with another solution that forms the cuticle on the surface. With its shell now in place, the egg is laid soon after.

Monotremes, like the duckbill platypus and echidna, show us that we are descended from egg-laying ancestors. Although they are mammals, they lay eggs containing early embryos similar in many ways to those of birds and reptiles.[15] They are large and yolky and have mineralized shells, and the amnion membrane forms from the fusion of head and tail folds that buckle up from the surface sheet of cells that forms in the first hours of development. Our eggs, in contrast, are now tiny and yolk-less, have no shell around them, and produce the amnion by formation of a cavity within a loose jumble of cells. Of course, this is possible because human (and other mammalian) eggs no longer have to support the embryo until birth—we use the placenta instead. The transition from egg laying to pregnancy in mammals is the subject of the next chapter.

STEP THREE: LOSE THE LARVA

Besides internal fertilization and a protective shell, amniotes made another major shift in their development: they eliminated the aquatic larval stage after hatching. Instead of a tadpole or other aquatic form, newborn hatchlings emerge as perfect miniatures of their parents, ready to live on land. This larva-free growth strategy is referred to as *direct development*.

Direct development greatly reduces mortality from dehydration but also from other hazards of the puddle, such as predators. Given these risks, committed amphibians living in nice, wet places would therefore seem also to benefit from ditching their larval form. Indeed, there are many recent losses of aquatic larvae within living amphibia. Some salamanders lay their eggs in moist places on the ground and guard them until they hatch into air-breathing juveniles. Frogs that live in trees have evolved large eggs that develop directly into tiny froglets in small pockets of water, such as bromeliad hearts or tree holes. In both cases, larval development was eliminated. To do so, they also evolved some curious novelties. For example, embryos of the beloved Puerto Rican tree frog, the coquí, develop their tails into a flat organ rich in capillaries. When pressed against the egg membrane, it functions in gas exchange, much like a mammalian placenta (figure 7.3).[16]

If direct development is possible and advantageous, why didn't all the amphibians join the amniotes in leaving the water behind? The answer is that loss of the larva comes with a catch: the egg must be provisioned with a much larger cache of yolk than that required to produce a simpler, smaller feeding larva. There is thus a trade-off—direct development has lower mortality, but it will reduce the number of eggs that can be made. In environments where standing water is present on a reliable basis, the ability to outsource much of the offspring's nutrition to the environment can allow the traditional, indirect strategy to produce more survivors and thus more descendants.

LOOKING AHEAD

In this chapter, we have seen how the allure of dry land pulled some amphibians out of the water. They achieved this by inventing penises to fertilize eggs without the sperm being shed into the environment, creating a water-conserving shell

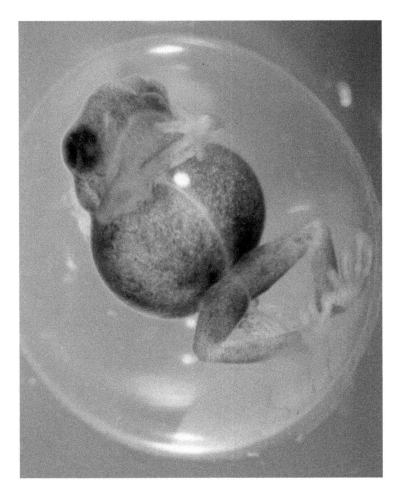

7.3 Direct development in the Puerto Rican coqui frog. The tail (*lower left*) of the Puerto Rican coqui frog has evolved into a flat, placenta-like organ for gas exchange. The young hatch from the egg with legs and lungs, skipping larval life entirely.
Source: Richard Eliinson.

and membranes around their embryos, and eliminating their aquatic larval stage. This cemented two features of sex and reproduction that we humans take for granted: that males have penises and that we don't have to put babies in swimming pools for their first two months of life. We also don't do the eggshell thing, but way back our ancestors did—we placental mammals just lost it as pregnancy evolved. How that happened is the subject of our next chapter.

CHAPTER 8

The Bun in the Oven

Synopsis

After laying eggs for tens of millions of years, some mammals evolved pregnancy to make reproduction more reliable. This transformation required both losses of the old and gain of new contraptions that were made from older parts. Live birth enabled mammals to adopt highly mobile lifestyles, to spread to the coldest places on Earth, and even return to the sea. It also set the stage for the dangers of human maternity.

———— ◆ ————

Let's geek out on beluga whales for a bit, shall we? They live in seawater so cold that it would freeze solid were it not for the salt dissolved in it. You and I would be dead after a few minutes immersed in their world, yet they are comfortable. Swaddled in thick blubber, they maintain a body temperature the same as our own. They fuel this typically mammalian physiology with a steady intake of fish, which they expertly hunt with the aid of a sophisticated echolocation system, much like the one bats use to hunt insects. To make their various clicks and squeals, they force air through special structures near their blowhole called (I kid not) the "monkey lips/dorsal bursae complex."[1] These various adaptations are impressive, and even more so because they evolved in the geological equivalent of a blink, over the last few tens of millions of years. But another crucial feature that enables their success came over 100 million years before their ancestors returned to the sea: they stopped laying eggs.

The platypus and the echidna give us a glimpse into how our ancestors made babies. They are furry and make milk but still lay eggs. The group of amniotes that includes modern mammals as its sole surviving lineage, the synapsids, laid eggs for something like 200 million years before pregnancy evolved. But once it did evolve, it allowed mammals to adopt lifestyles that would have been impossible otherwise. A beluga whale cannot sit on a nest (heck, it can't sit anywhere!), it has nowhere to build it even if it could sit, and the water is too cold to keep the egg warm anyway. Pregnancy allowed mammals to spread into new habitats on land as well. Think of antelope and other grazers that are constantly on the move, and monkeys that live their entire lives moving from tree to tree. Pregnancy means being able to go wherever food is abundant and predators are scarce without having to incubate and protect eggs in a fixed nest. By being able to bring the developing offspring along, new lifestyles became possible, lifestyles that would eventually include our own.

WHEN?

Mammalian pregnancy evolved earlier than you might think. Recently discovered Chinese fossils reveal that the marsupial and placental mammals (the *metatherians* and *eutherians*, respectively) had already become distinct from their egg-laying relatives in the Jurassic, at least 160 million years ago, and diversified further in the Cretaceous (figure 8.1). A third major group of mammals, the multituberculates, was also expanding in the Jurassic and Cretaceous. You won't find them in the zoo, however: having survived the end-Cretaceous impact that wiped out the big dinosaurs and much of the rest of life on Earth, multituberculates went extinct roughly 30 million years later.[2]

In 1970, pioneering Polish paleontologist Zofia Kielan-Jaworowska demonstrated that Cretaceous multituberculates had pelvises far too narrow to allow passage of either a shelled egg or live offspring at an advanced stage in development. She concluded they were live bearers whose newborns were tiny, much as those of marsupials are today.[3] If this also applied to Jurassic multituberculates, then we cannot reject a single, very ancient origin of mammalian pregnancy. It's also possible, however, that it evolved more than once. A recent comparison of the bones of the three major mammalian groups suggests that baby mutituberculates grew more like placentals than marsupials.[4] This surprising result means surviving groups only exhibit a few of a historically more diverse set of reproductive strategies used by early mammals.

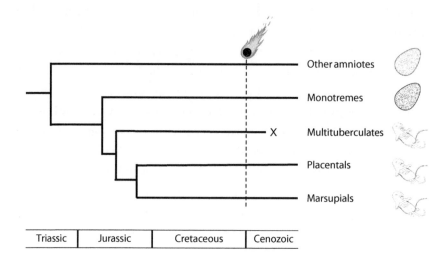

8.1 Major mammalian lineages, with their reproductive mode (oviparity via laid eggs or viviparity via live birth) indicated at the right. The dashed line marks the end-Cretaceous mass extinction 66 million years ago.

Source: Inset drawings by Louisa Wu.

We think of the Jurassic and Cretaceous Periods as dinosaur time, and it surely was. Yet the three large groups of mammals that no longer laid eggs were already diversifying and evolving new ways of life. Some lived on the ground, while others were specialized for climbing trees. By the time the big dinosaurs were wiped out at the end of the Cretaceous (with birds as the only survivors), there were many mammals ready to jump into the gap. Pregnancy likely helped live-bearing mammals to invade new habitats rapidly.

Early mammalian diversification occurred against a backdrop of changing plant life. Jurassic forests included tree ferns and nonflowering seed plants such as conifers, cycads, yews, and ginkos. Later, in the Cretaceous, flowering plants (*angiosperms*) rapidly radiated. Trees created a new kind of habitat, one offering protection from ground-dwelling predators, as well as insects, leaves, seeds, and fruits as food. While most early mammals, including the oldest known multituberculate,[5] lived on the ground, spectacular fossil discoveries from the last two decades have revealed that the first known marsupials and placentals were adapted to life in trees.[6] Many of these discoveries were made by paleontologist Zhe-Shi Luo, currently at the University of Chicago, and his colleagues. In 2011,

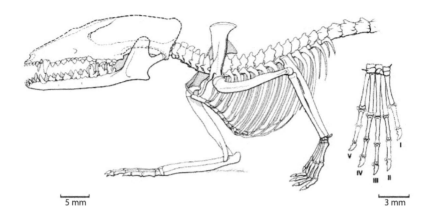

8.2 *Juramaia sinensis*, the oldest known placental mammal. It lived 160 million years ago in what is now Liaoning Province, northwestern China, and had hands specialized for climbing (*inset at right*).

Source: Z. X. Luo, C. X. Yuan, Q. J. Meng, and Q. Ji, "A Jurassic Eutherian Mammal and Divergence of Marsupials and Placentals," *Nature* 476, no. 7361 (August 2011): 442–445, https://doi.org/10.1038/nature10291.

Luo's group described the late Jurassic placental, *Juramaia sinensis*. About the size of a shrew, it has skeletal features found today only in tree-dwelling mammals, such as lemurs (figure 8.2).

WHY?

As handy as live birth eventually proved to be, evolution does not have foresight. Any innovation that had to wait until the demise of the large dinosaurs before it could be advantageous would never have evolved in the first place. So what was the *immediate* benefit to those early mammals that evolved it? To be honest, we don't yet know. One can imagine many reasons why egg laying might become less reliable, such as hungry nest predators, or limits to how far parents could search for food imposed by having a nest of eggs. If live birth in mammals was a truly unique historical event—an ancient singularity—we lack the replicated cases that can reveal the underlying cause of a transition. Is there anything else we can bring to bear?

If we need multiple origins of pregnancy in amniotes, there is a different group that can help us: snakes and lizards (the squamates). They have evolved extended fetal development many times, most supported by elaborate placentas with remarkable (but totally convergent) similarity to those of placental mammals. Why some squamates evolve live birth has been a much-debated topic among herpetologists. As the underlying causes are relevant to our own mammalian ancestors, a digression is in order.

Early twentieth-century zoologists working in spots as distant as Mexico, China, Australia, and Russia all noted the prevalence of live birth in species living in colder habitats. The work of this cast of pioneers and their successors was codified as the cold-climate hypothesis.[7] The basic idea was that retention of the embryo in the mother shielded it from cold. While lizard eggs can actually tolerate brief exposure to very cold temperatures, the faster development imparted by internal development may be the real benefit here.[8]

The ability of prolonged gestation to prevent fetal chilling would be even more true for a warm-blooded mammal. However, the Jurassic was an unusually warm time on Earth, so invoking the squamate cold-climate hypothesis for our own ancestors seems a bit of a stretch. Even for lizards and snakes, the story is not that simple. Noted herpetologist Richard Shine has argued that there are benefits of internal gestation even without the cold, all revolving around the increased maternal control over the conditions of fetal and neonatal life achieved by live birth. This broader benefit of pregnancy has been dubbed the maternal manipulation hypothesis.[9]

The most ancient placental and marsupial animals apparently lived in trees. Could this have provided another impetus for increased maternal manipulation? Daniel Blackburn of Trinity College in Hartford, Connecticut, has spent his career studying the many independent origins of live birth in squamates. One of Blackburn's recent studies, conducted with Daniel Hughes of the University of Illinois, is particularly relevant here. Hughes and Blackburn found that, in the chameleons, a group of over two hundred species, live birth has evolved three distinct times over the last 50 million years. In each of these cases, it was preceded by adoption of an arboreal lifestyle. Thus, the move to trees somehow predisposes them to switching reproductive mode.

Just as for cold, however, a tree-dwelling lifestyle is insufficient to explain why squamates give up laying eggs. As Blackburn explained to me, "many arboreal squamates do just fine with oviparity—[they] just come to the ground to lay eggs." Each case of live birth in tree-dwelling chameleons "occurred in a high altitude lineage whose species experience seasonally cold climates." Some fully ground-dwelling lizards, such as North American horned lizards, have also evolved viviparity in higher, cooler places even without living in trees.[10] So the best predictor for lizards is the *combination* of trees and cold, maybe because the additional exposure to wind and cold above the ground makes egg laying too risky.

I asked Lucas Weaver, an expert on early mammals at the University of Michigan, why he thought live birth evolved, and specifically how relevant cold and arboreal life might have been. He was skeptical: "[Live birth] almost certainly arose sometime in the Late Triassic or Jurassic, a 'hothouse' climate in terms of Earth's global temperature through time. As such, the differences in temperatures on the ground versus in the trees would likely not be a sufficient driver of viviparity. To me, viviparity being a consequence of more intimate contact between mother and offspring seems a more likely driver—if juvenile survival was so tightly tied to the mother supplying nutrients, then I could see selection favoring a reproductive system that would ensure the mother was immediately on hand to nurse the young as soon as they were born." This sounds more like Shine's maternal manipulation hypothesis.

An admittedly speculative scenario that pulls together what is known about the origins of mammalian pregnancy goes something like this. Because living, egg-laying mammals and most tree lizards lay their eggs in ground nests, it's likely that the first tree-dwelling mammals did so as well. During their reproductive season, they commuted back and forth between ground nests and food sources in the trees. This worked well enough for a long time, but something changed that incentivized increased maternal control over gestation, the time and place of birth, and/or the feeding of young. Perhaps foraging trips into the trees grew longer, allowing the eggs to be discovered by nest predators. Digging a burrow could help, but tree dwellers' bodies are specialized for climbing, not digging. Burrows also anchor the family to a spot for months, preventing movement to better sites. Pregnancy allowed the gestating female to relocate at will, to speed the development of her offspring, and to reduce their risk of being eaten by nest predators. It also let her gradually allocate the resources required to support her babies' development instead of providing it all up-front in a big egg. Any or all of these benefits may have caused our ancestors to bid adieu to egg laying.

Once pregnancy evolves in an amniote, it never reverts to egg laying. This sort of one-way evolutionary street is referred to as Dollo's law, after the Belgian biologist Louis Dollo, who apparently first formalized it.[11] Dollo's law is especially apparent in the evolution of reproduction—and not only in amniotes that lost egg laying. For example, some frogs and sea urchins, which historically had extended larval development and dramatic metamorphosis, can compress their life cycle by skipping the larval stage.[12] This is usually accompanied by a big increase in egg size and a big decrease in egg number. As with live birth, there are no clear cases of reversion to typical larval development.

Such irreversibility might be because the new invention is just so awesome that nothing can improve upon it. But as we will see next, it is also because the key tissues and genes that nourish and protected the egg were either radically altered or lost entirely. Once gone, there is no facile way to restore them.[13] There was no going back, and mammalian history would be altered forever.

HOW?

While human females don't lay a watermelon-sized egg in a crib (whew!), our ancestors more or less did the equivalent. How did they manage to transition from laying eggs to pregnancy? If we compare a chicken to ourselves, we can see a number of changes that had to be made. An obvious one is the egg's shell, which we examined in the previous chapter. This amazing invention was the amniote's ticket out of the swamp and mediates a delicate balance between water retention, contamination prevention, and gas exchange. Yet it was mostly or completely lost in lizards, snakes, and mammals that have live birth.

Did the shell have to be lost? Why couldn't a fertilized egg simply be held in the mother until shortly before hatching without any other changes? This is called ovoviviparity and is fairly common in lizards and snakes. The reason mammals don't do this likely has to do with metabolism and breathing. At some point in its development, a warm-blooded embryo would reach a level of metabolism at which it would need to gain more oxygen and give off more carbon dioxide than an internal, shelled egg could manage. By allowing an intimate interface between the maternal and fetal blood supplies (the placenta), losing the shell made the import of oxygen and export of carbon dioxide much more efficient.

Another obvious feature of a chicken egg we lack is the yolk. Amniote eggs that are laid must provide all the nutrition the baby will need to develop and hatch in a completely closed system. Even with the eggshell gone, this strategy could continue inside the female parent and probably did for some time. But eventually the uterus-placenta interface allowed nutrients imported from maternal blood to replace the yolk. In the process, membranes surrounding the embryo (the allantois, amnion, and chorion) were repurposed to form the placenta, which became the site for the exchange of gases, nutrients, and waste products with the mother.[14] So live birth is far more than losing the shell—it required elaborate new inventions as well. The before and after is illustrated in figure 8.3.

The previous two paragraphs summarize what happened as mammalian pregnancy evolved, but they still leave out the details of how it happened. By that, I mean the details of how genes, proteins, cells, and their interactions in development were modified. I have long been fascinated by such details because, in evolution, they are the place where the proverbial rubber meets the road. The most helpful adaptation, like, say, pregnancy, cannot spread by natural selection until it actually exists in some individual. This mysterious process was dubbed "the arrival of the fittest" by Jacob Gould Schurman (1854–1942), a Canadian-American philosopher, university president, and diplomat. Schurman accepted evolution, and his main interest was in justifying ethics in an increasingly materialistic world. In 1893, however, he correctly noted that the tendency of organisms to produce novel forms spontaneously was still unexplained by Charles Darwin and implied it was so improbable as potentially to require divine origins.[15] Biological novelty is indeed a kind of miracle, but the modern field of evolutionary developmental biology (evo-devo) has begun to explain it in terms of specific mutations having an impact on genes, proteins, and the regulatory circuits they form.[16]

The fact that the genes that regulate development are key is obvious upon a bit of reflection: muskrat mammas give birth to baby muskrats and not to baby sea urchins. This predictability comes from the genome—muskrat DNA contains the instructions for making muskrats and not anything else. It follows therefore that any long-lasting change in the form of an embryonic or adult organism needs to somehow involve modifications of the genome. In addition, organisms cannot take a few generations off from the business of surviving to solve a problem; they must be making these modifications while maintaining all other essential processes. It's a bit like driving nonstop from New York City to Denver, Colorado, in a vehicle that begins the trip as a pickup truck but somehow arrives 1,800 miles (2,800 km) later as a two-seater convertible. This is a real form of biological magic and never ceases to astound me. When my developmental

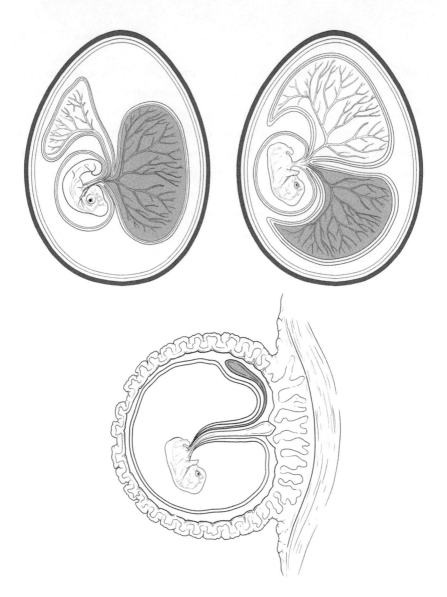

8.3 Transformation of the embryonic membranes in mammalian pregnancy. Comparable stages of development are shown for a bird (*upper left*), monotreme oviparous mammal (*upper right*), and human (*bottom*). In each case the outermost membrane is the chorion, the innermost membrane surrounding the embryo is the amnion, and the yolk sac is shaded in gray. In the oviparous amniotes (*top*), the allantois is a large sac connected to the hindgut, which assists in gas exchange and also acts as a sort of diaper. It remains in humans as a small vestige imbedded in the umbilical cord. The human chorion is highly convoluted, especially where it forms the interface with the uterine wall at the right. This creates lots of surface area for exchange of nutrients, gases, and wastes with the maternal blood.

Source: Nicholas Bezio.

biology professor at Oberlin College, Yolanda Cruz, first explained how the evolution of development brings new traits into existence, it was like being zapped in the head by a lightning bolt. It implied that, with genome sequences and some experiments, it might actually be possible to reconstruct the specific changes that allowed new things to evolve—amazing new things like pregnancy. Below is a collection of insights on the evo-devo of mammalian pregnancy gathered by my colleagues over the last twenty years.

Coming Out of Our Shell

As noted above, the loss of the eggshell must have been one of the first steps in the evolution of pregnancy. The simplest idea for how that occurred would be that the genes responsible for building the shell were simply deleted from the genome of our ancestors, but this turns out not to be the case.[17] This is because many genes that are especially active in the shell gland of egg-laying amniotes encode proteins involved in depositing calcium carbonate and calcium phosphate, and the eggshell is not the only place in the body they are active: our bones are also composed of calcium compounds. Because of this requirement, "eggshell genes" are still found in mammals that don't lay eggs; however, they are turned off in the uterus, the mammalian counterpart of the shell gland. This kind of tissue-specific change in a gene's activity has emerged as one of the most important ways evolution modifies development. For example, my University of Maryland colleague Sean B. Carroll and his lab have documented similar changes that lead to the appearance of spots on the wings of fruit flies whose recent ancestors had none.[18]

While losing the shell around the egg allowed the placenta to evolve, it created another problem: the eggshell is the historical source of calcium for bone growth, so a new way of getting calcium to the fetus was needed. For humans, the need is especially great in the third trimester, when bones grow rapidly and 80 percent of a newborn's calcium is acquired. The key protein that imports calcium into the fetal bloodstream, TRPV6, is similarly active in birds and mammals, so that is unchanged.[19] Instead, the innovation was to increase the calcium available in the maternal blood during pregnancy by boosting the absorption of calcium in the mother's intestine.[20] This in turn is driven by the production of more vitamin D.[21] Once again, an important difference between egg-laying and live-bearing mammals is explained not by genes being gained or lost but rather by changes in their activity.

That Ain't No Yolk

Besides losing their shells, placental mammal eggs also lost over 99.9 percent of their volume, as well as the gooey white, or albumen, that swaddled them in the past. In contrast to shell loss, these changes involved inactivation and eventual deletion of genes from the genome.

The yolk proteins used in eggs across all animals, the vitellogenins, are recognizably similar from shrimp to sharks. Because their main job is to be digested, this seems surprising at first. The conservation of vitellogenin in yolk over hundreds of millions of years relates to its ability to bind the kind of fats that are used to make up cell membranes. Embryos use yolk to provide the membranes around their many new cells without eating. Because the fats would otherwise be insoluble and sticky, vitellogenin allows them to move from the liver, where they are made, to the growing eggs in the ovary. Related proteins chaperone fats in other parts of the body, including the high-density lipoproteins (HDLs, or "good cholesterol") your doctor may be monitoring to keep your heart healthy. The fat-ferrying job of vitellogenins is so crucial for embryos that it has kept them around since the Cambrian Explosion.

One group of animals where you won't find vitellogenins, however, is that comprised of live-bearing mammals. Both marsupials and placental mammals like us have no active vitellogenin genes left, although mutated, nonfunctional copies called pseudogenes can still be found. When did our ancestors' yolk genes get wiped out? Did it happen all at once? The eggs themselves offer some clues: platypus eggs are less yolky than those of birds, suggesting that nursing may have allowed some loss prior to the debut of live birth. Second, despite using live birth, marsupial eggs retain a small amount of yolk-like material in their eggs, although its exact composition remains unknown.[22] All these factors indicate that the loss of yolk on the road to human-style pregnancy was probably gradual.

Molecular fossils in the genome also help us understand the loss of yolk. To find them, we need genome sequences from both egg-laying and live-bearing mammals, plus some other egg-laying amniotes for context (for example, birds). In 2008, a team of Swiss biologists at the University of Lausanne, led by David Brawand, pulled this all together to reconstruct the history of the vitellogenin genes in mammals.[23] What they found is fascinating and largely consistent with what was already known about the amount of yolk in eggs. While birds and frogs each have three vitellogenin genes, the platypus has only one (plus a recently inactivated pseudogene). Placental mammals not only lack obvious yolk in their eggs, they have also lost all functional vitellogenin genes. Despite appearances of yolk in their eggs, marsupials also have none.

Genes that code for proteins of the egg white have also been lost. For example, the one responsible for the stiff peaks in your meringue is ovalbumin, which is encoded by three highly similar genes that sit next to each other in the chicken genome. Turtles and alligators have similar proteins in their own eggs. If we look in the human genome, however, we find the instructions for making related proteins used in other parts of the body but no exact counterpart of the egg ovalbumin genes. This all suggests mammals used to have an egg-specific ovalbumin like other egg-laying amniotes but lost it.[24]

Besides ovalbumin, another egg white protein has been lost in us mammals: riboflavin-binding protein (RBP). RBP's job is to shepherd vitamin B2 from the female's somatic tissues to the growing egg cell. RBP is still used by the platypus, but its job became unnecessary in placental and marsupial mammals once the maternal and fetal bloodstreams were able to contact each other in the placenta and early growth became supplemented by milk. As a result, neither group of live-bearing mammals retains a functional copy of the gene encoding RBP.

Building the Placenta

Because our eggs no longer have shells or yolk, they are very small (about a tenth of a millimeter across) and squishy. The sperm finds and fertilizes the egg in the tube connecting the ovary and uterus (called the *fallopian tube*), and the zygote begins to divide. By the time it reaches the uterus, it has become a fluid-filled ball of cells, the *blastocyst*. The blastocyst snuggles into the soft, blood-rich lining of the uterus and soon forms a strong connection with it. The outer layer of the ball produces only extraembryonic membranes and the placenta, while the inner part gives rise to the fetus itself.

I think it's fair to call the mammalian placenta a Rube Goldberg contraption. It is built using tissues and genes that had other functions when we were egg layers. In addition, once the basic problem of pregnancy was solved, evolution continued to tinker with the details. For example, marsupials and rodents mainly use the yolk sac to begin building the placenta, and they continue to use the allantois as a sort of embryonic diaper to collect wastes. In contrast, we humans use the allantois to form the first connection with the placenta and only incorporate parts of the yolk sac later as a source of blood supply (figure 8.3).[25]

The miracle of the placenta can also be appreciated from the standpoint of immunology. The rejection of organ transplants shows that even tissues from close relatives are still recognized by our immune system as "not us," and they are attacked as if they were malicious invaders. The beginning of pregnancy is actually a very similar situation. Genetic variants not shared by both parents lead to embryos that are partially foreign from the perspective of the mother's immune system. Yale biologist Günter Wagner and his colleagues have shown that embryo implantation in the uterus wall triggers immune inflammation typical of an infection.[26]

The uterine inflammation induced by the embryo is initially helpful because it is accompanied by concentration of maternal blood to the site of implantation. This is essential for a placenta to function. If left unchecked, however, the immune response advances to a point where the embryo would be killed. Thus, to use this response productively, it is crucial that the initial response be frozen in place. Marsupials and placental mammals both show early inflammation, which tells us that it was a feature of their common ancestor. But they differ in how they manage it. In placental mammals, a durable placenta is formed that connects the mother and fetus for over a year in some cases. In contrast, marsupials only delay the immune response briefly so that their offspring's placenta is rejected like a failed organ transplant after only a few weeks.[27] This triggers the birth of their characteristically immature, tiny babies, which finish development in the pouch.

The inflammation aspect of placental development is interesting but by no means the weirdest part of the Rube Goldberg contraption. Another important feature of the placenta is made possible by an actual foreign invader, one that was domesticated and put to work. The invader is a retrovirus, an infectious agent that can insert itself into our DNA after passing through an RNA intermediate. HIV, for example, is a retrovirus we try to avoid. However, our genomes have been bombarded by many retroviruses over millions of years, and they have accumulated. Most are inactive molecular fossils, and together they make up about 5 percent of our genome, a major component of our "junk DNA." However, in 2000, it was discovered that one of these old retroviruses, HERV-W, plays a crucial role in the development of the human placenta.[28]

How could a retrovirus possibly help the placenta grow? The fetus-derived cells of the placenta at the interface with the uterus, called *syncytiotrophoblasts*, have multiple nuclei. They get that way because typical, single-nucleus cells fuse with each other. This fusion creates a gap-free tissue at the boundary that prevents maternal immune cells from invading the placenta, which would be a disaster for the fetus. It is this cell fusion step where HERV-W comes in. A key moment in a retroviral life cycle is when a free virus's membrane jacket fuses with the membrane of the cell it is infecting. Because this fusion step is essential for the virus, the proteins that enable it, called *syncytins*, are really good at their job. HERV-W syncytins are massively produced in syncytiotrophoblasts and their single-cell precursors, and they can be used to induce fusion of just about any animals cells that also have a receptor protein on their surface.

The domestication of retroviral syncytins by the placenta is an amazing example of how evolution shamelessly takes advantage of whatever is on hand. But opportunism is not the same as simplicity: Getting syncytins to be produced in precisely one cell type of the placenta required rare changes to the DNA surrounding the retrovirus copy. This constraint would lead one to expect that this was a singular event in mammalian history, like the internalization of the mitochondrion by our archaeal ancestors discussed in chapter 1. So all placental mammals with syncytiotrophoblasts use syncytin from HERV-W, right? Wrong! Different mammals use different retroviral sources for their syncytins. For example, carnivores, like cats and bears, began to use a syncytin from around 85 million years ago and have stuck with it since then. Sheep, rabbits, and rodents use still others.[29] Perhaps most amazing is a lizard that recently evolved live birth and a placenta, whose syncytial cells also captured a retroviral syncytin.[30] It is likely that the sheer number of retroviral copies in vertebrate genomes make it not so hard to press them into service in the placenta after all.

Hatching

In our fish and amphibian ancestors, the embryo was protected by a thin protein membrane until it was old enough to swim. At that point, the hatchling-to-be would break down the membrane by secreting a protein called hatching enzyme and then swim away. Amniotes retain this hatching event, but its details necessarily evolved. Bird eggs also have a protein membrane around them—you rupture it every time you crack an egg. As with fish, it lingers through development, but it is eliminated before hatching out of the shell using a fish-like hatching enzyme.[31] In other words, bird embryos actually hatch twice: once from the ancient membrane around the egg and a second time from the shell. What about live-bearing mammals? Shedding of the egg membrane must happen before the embryo implants in its mother's uterus at the blastocyst stage. This is very early in development—long before any recognizable body parts have formed. Nevertheless, hatching is induced by the same enzyme used in fish and birds. This extreme shift in timing of hatching enzyme production was one important step of the many on the road to pregnancy.

THE SILENT TWIN

A few minutes after my wife delivered our first daughter into the world, out came the placenta that had supported our baby for the previous nine months. Weighing over a pound, it was a deep red blob that reminded me of liver. Its highly branched blood vessels were obvious, and strangely fascinating. These blood vessels create the tremendous surface area needed to support the exchange between mother and baby—although the placenta is only 10 inches (25 cm) across, it has been estimated to have a total surface area of 12 to 14 square meters (123 to 150 square feet). After being securely attached to the uterus for nine months, it detached in a matter of minutes—basically as soon as it was no longer needed. This precisely timed ejection was honed by tens of millions of years of natural selection acting on mammalian reproduction. I was so amazed I actually took a picture, although I'll spare you the grisly image here.

   I am far from the first mammal to marvel at the afterbirth. From jackrabbits to jaguars, mothers eat the placenta to recycle the nutrients it contains. While humans are unlikely (but not unheard of) to do that, the placenta is nevertheless treated with reverence and given a ceremonial burial in many cultures. In the Bergama region of Turkey, for example, the placenta was regarded as a companion or sibling of the newborn baby and was buried.[32] For Maoris of New Zealand, the place where the placenta is buried helps bind the newborn child to its ancestral homeland. In contemporary life, some Maori mothers who give birth abroad send (or personally escort) the placenta home for burial, a practice facilitated by cooperation from the New Zealand Customs Service.[33]

THE DOWNSIDES OF LIVE BIRTH

I hope that you now have a new appreciation of the wonder that is pregnancy. As you may know from firsthand experience, however, it also poses risks for baby and mother alike. The diciest times for a pregnancy are a bit like those associated with airplane flights: at the beginning and the end. Eighty percent of known miscarriages occur in the first trimester, and those not due to genetic defects often stem from a failure to sustain the placenta. Abnormalities of the uterus, premature opening of the cervix (the constricted opening that holds the fetus in until delivery), or infections can all derail a pregnancy. If a robust placenta is formed, however, the middle period of gestation has a relatively low risk of failure.

As the fetus nears its due date, the dangers of childbirth loom. Preterm birth leaves newborns at a high risk of death from respiratory failure. This claimed my oldest sister, Wendy, and almost got me. For full-term births, the baby must manage to squeeze through a birth canal that is just barely big enough for it (figure 8.4). Even for a perfectly healthy fetus, birth can fail if it has an abnormal body position during labor (breach birth). Newborns can also be strangled by

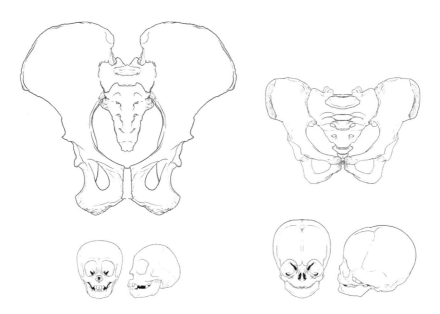

8.4 Comparison of the fetal head and the female pelvis in humans and gorillas. The female gorilla (*upper left*) has a larger pelvis overall, and the neonatal gorilla head (*lower left*) fits comfortably through it. The human female pelvis (*upper right*) is smaller overall, but its birth canal is similar in width to that of the gorilla. The much larger human neonatal skull (*lower right*) is thus a tight fit.
Source: Nicholas Bezio.

the umbilical cord during birth or deprived of oxygen if the placenta becomes infected during a prolonged labor.

The risk of a human baby dying during labor is, at present, highly dependent on the availability of obstetric care. While one baby in three thousand dies during birth in Ireland, the rate is roughly one hundred times that in Tanzania. This higher rate is likely to be closer to that experienced by humans throughout most of our history. These costs are real and, of course, wouldn't exist if we still laid eggs like birds or platypuses. But there is no going back—too many essential features of shelled egg production have been lost.

The baby is not the only one at risk during pregnancy and childbirth, and this is especially true for our own species. In developing countries without reliable obstetric care, the risk is staggering. In South Asia, nearly 2 percent of women will die because of complications of one of their pregnancies or labors, and in sub-Saharan Africa, the rate is over 4 percent.[34] Although they do reflect intrinsic risks of human pregnancy and childbirth, in the twenty-first century, most of these deaths are needless. The United Nations Human Rights Commission put it this way:

> The majority of maternal deaths and disabilities could be prevented through access to sufficient care during pregnancy and delivery and effective interventions. This affirmation is supported by the observation that in some countries maternal mortality has been virtually eliminated. Only 15 per cent of pregnancies and childbirths need emergency obstetric care because of complications that are difficult to predict. The WHO [World Health Organization] estimates that 88 to 98 per cent of maternal deaths are preventable. More recently, UNICEF has reaffirmed that approximately 80 per cent of maternal deaths could be averted if women had access to essential maternity and basic health-care services."[35]

Pregnancy is so risky for humans because it pushes their bodies to the absolute limit of what is physically possible. During gestation, the fetus's demand for nutrients can elevate maternal blood sugar, causing gestational diabetes. The need to dampen immune responses in the uterus, which prevents rejection of the placenta, also places the mother at risk for bizarre opportunistic infections. For example, the bacterium *Listeria* can attack pregnant people if they eat foods that would be benign at other times of life (or to their mates), such as soft cheeses or cured meats. Similarly, exposure to dirty cat litter can infect someone who is pregnant with *Toxoplasma*, a tiny eukaryotic parasite related to malaria that pets can pick up if they spend time outside. Imperfections in the placenta-uterus interface can also release proteins that harm the mother's blood vessels, leading to preeclampsia, a condition causing life-threatening high blood pressure and kidney damage.[36]

Even if a gestating female avoids the veritable minefield of complications above for nine months, she is not out of the woods yet. The head of a newborn human is really big compared to the birth canal. As a result, the mother can suffer from a blocked delivery or tearing and bleeding, either of which can be lethal without intervention and sanitary recovery conditions. Most mammals are not at such a high risk of dying during birth and typically deliver multiple babies without a problem. Not even other great apes, our closest kin, have such difficulty with birth, as can be seen by comparing the human and gorilla pelvises and newborn head sizes (figure 8.4). Why do we have it so bad? If so many of us die from labor, why didn't natural selection produce a more accommodating pelvis?

Physical anthropologists have long blamed two innovations of our species that set us apart from other primates—extreme intelligence and upright walking—for the ardors of childbirth. In 1960, Sherwood Washburn proposed that the "obstetrical dilemma" of a big head in a tight pelvis represents a compromise between the two: the head needs to be big to hold the big brain, yet the pelvis must remain narrow to make the equally important daily task of walking efficient. Any change to address one need would antagonize the other.[37] Washburn also suggested that the demands of labor pushed human fetuses to be born at a relatively immature state so that their heads are still small enough to fit, but unlike other primates, our newborns cannot hold onto their parents. The need to both nurse and carry babies for a least a year constrained parents' other activities in new ways, especially for the mother.

I first learned about the Washburn hypothesis in graduate school in a lecture by anthropologist Owen Lovejoy, and the argument seemed perfectly plausible. However, in the last few years, researchers have pointed out that not everything aligns with it. For starters, it predicts that a wider pelvis should harm walking or running performance relative to a narrow one, all else being equal. When Anna Warrener, of Harvard and the University of Colorado, Denver, examined the evidence in 2015,[38] she found it didn't hold up in two ways. First, the idea that a wider pelvis inhibits performance is based on a simple, pendulum-like model of the forces acting on our hip joints. A more detailed model that takes all the motions of the legs into account makes this less clear. Second, when Warrener actually measured the per-pound energy required by people to walk and run on a treadmill, she found no relationship between it and the width of the pelvis *within* a sex. And while the key muscle predicted by the simple model to work harder when the pelvis is wider (the hip abductor) does indeed work harder, it actually contributes relatively little to the cost of walking and running. As a result, there is no difference between male and female people in that cost.

Washburn's assertion that the helplessness of human infants is due to the need to pass through the birth canal before the head gets too big has also been challenged. University of Rhode Island anthropologist Holly Dunsworth has shown that human babies are born at the time predicted to be the point at which the metabolic demands they make on a gestating parent become unsustainably great.[39] Thus, even if a wider pelvis had evolved, it probably wouldn't have resulted in babies being born later—it is already at the limit for metabolic support. Why human babies evolved to be unable hold on to their mothers like other primates is thus still a mystery.

So if a wider pelvis doesn't make walking harder, and being only a little wider would make birth relatively routine, then we are back to the question of why *didn't* it get wider. An alternative explanation for why pelvic width is constrained was put forward by Mihaela Pavlicev and her colleagues.[40] They argue that there are other benefits of a narrow pelvis that Washburn had not considered, including the need to support the weight of internal organs and, during pregnancy, the ever-heavier fetus. A follow-up study confirmed that a wider pelvis would create serious problems independent of walking.[41] Selection on pelvic width also affects males. Before puberty, both sexes have similar pelvic shapes, but young women diverge from the common plan under the influence of sex hormones (see chapter 10). This has its limits, and more extreme widening for females would affect both sexes. In particular, Pavlicev notes that the same pelvic floor muscles that can weaken during pregnancy to cause incontinence are crucial in males for a rather different function: erection. To put it a bit crudely, no boners, no babies.

GENETICS AND PREGNANCY

Chapter 5 noted that turtles and alligators use temperature to determine their sexes, while birds and mammals use chromosomal differences instead. Because pregnancy holds the fetus at a fairly constant maternal temperature, the use of chromosomes in our ancestors would have had to originate no later than the time when live birth evolved. This can be tested by looking at egg-laying mammals, and the results were surprising. In 2004, Jennifer Marshall Graves and her colleagues reported that the platypus has a unique form of chromosomal sex determination.[42] Instead of a male having one X and one Y like other mammals, they have five X's and five Y's, while the female has ten X's. Marshall Graves's team found that this enlarged set is accurately sorted every generation because each sex chromosome partially overlaps with another of the same type (that is, the X's overlap with other X's, and the Y's with other Y's). This forms a bizarre chromosomal conga line during meiosis. Odd though this may be, the existence of sex chromosomes in the platypus tells us that genetic sex determination likely predated the origin of live birth.

Then the platypus thing got even weirder. After the discovery of the crucial male-determining *SRY* gene in the 1990s, it became synonymous with the function of the mammalian Y chromosome. So one of the five platypus Y's must harbor a copy of *SRY*, right? Wrong again! Meticulous searches of both the platypus and echidna genomes have found no Y-linked *SRY*-like gene.[43] In addition, the genes on the platypus X's and Y's are not generally found on our X and Y. They are shared to a large extent, however, by the distinct sex chromosomes of birds (called Z and W). We can thus disconnect the general phenomenon of genetic sex determination in mammals, which appears to be very ancient, from the specific use of the *SRY* gene as the deciding upstream factor, which must have arisen around the time of live birth. Whether this is a coincidence or the two are connected in an important way remains unclear.

Pregnancy is associated with other genetic quirks. Egg-laying mammals nourish their offspring early via the yolk and after hatching through nursing. Neither of these ways of transferring materials can be manipulated by the growing baby—the mother is in control. But when pregnancy evolved, material transfer became a two-way street mediated by the placenta. The fetus could thus demand more nutrients by enlarging the size of its placenta or manipulating the physiology of the mother. Males competing with other males stand to benefit the most from this because their offspring would be relatively bigger. Countermeasures would need to evolve to keep this in check lest the mother's other offspring and/or her own health suffer from greedy progeny.

A phenomenon called genomic imprinting indicates such a "war of the womb" between parents actually happened early in placental mammal evolution. In an embryo, the two copies of each chromosome are not always treating equally. The first indications of imprinting were discovered in the 1970s, when researchers learned that the placenta of XX mice and humans (but not the embryo itself) selectively shuts off the X chromosome derived from the sperm genome. While one X is also inactivated in the embryo proper to match the single X of the male, in that case the parent of origin is randomly chosen. This indicates there is something dangerous about allowing the sire's X to remain active in the placenta. Marsupials go further, and shut off the paternal X everywhere.

Parent-of-origin effects on fetal gene activity are not restricted to the X.[44] Forcing a mouse embryo to use two maternal or two paternal genomes is lethal, even if the embryo is XY (and thus has only one X). The 1990s saw an explosion of discoveries of imprinting in autosomal (i.e., not X-linked) genes. The general pattern is that the sperm genome is trying to stimulate fetal growth, while the oocyte genome is imprinted to limit it. Imprinting can only turn genes off, so each parent inactivates genes to oppose the tendency of the opposite sex to hyperactivate the same genes. Thus, oocytes imprint the maternal copy of a growth-promoting gene that is paternally active, while sperm imprint their copy of genes that are maternally active and repress growth. Imprinting thus appears to represent a frozen stalemate between the opposing parental interests. The fact that this is directly related to pregnancy is confirmed by egg-laying amniotes, which have the same genes but do not imprint either parental version.

CONCLUSION

Pregnancy requires many modifications of our egg-laying ancestors, some quite bizarre. It has been a feature of our lineage for at least 160 million years, a time of amazing success for placental mammals. We have spread to every part of the world, from the icy shores of Greenland to the deep ocean, from rainforests to the driest deserts. In many of these habitats, egg laying would have prevented reproduction, so it is fair to say pregnancy was a crucial enabler of our success.

Our own species is especially vulnerable to childbirth complications, yet we are closing in on 8 billion people. Clearly pregnancy works well enough for us, but it also imposed extra burdens on female parents that linger in modern society. Their amelioration requires both their recognition and deliberate planning, issues I take up in the final chapter.

PART III

Sexual Homo

CHAPTER 9

Period Piece

Synopsis

The monthly fertility cycle of human females evolved from a system in which egg release was triggered by the act of mating. Ovulation independent of mating was followed by anticipatory maturation of the entire uterus. This costly investment evolved as a countermeasure to protect the mother against increasingly invasive placentas. Menstruation saves energy in cases where no pregnancy occurs. Although it can impose a physiological burden, irrational taboos about it put women at even greater disadvantage.

My wife, Shizuka, often lets me know how annoying menstruation is. "Who would ever design such a stupid system?" As she hit fifty, she expressed the monthly hope that maybe this was her last period. I have never had to suffer through it, which seems grossly unfair. Since I'm supposed to know something about the evolution of reproduction, her question has stuck with me. In solidarity with her and with all who menstruate, this chapter is my attempt to give her and others a decent answer, although much remains mysterious.

First, what exactly is menstruation? It is the shedding of the entire, ripened inner layer of the uterus. This layer is called the *decidua*, related to "deciduous," which you hear referring to trees that drop their leaves. In Latin, *decidere* means "to cut off," and the decidua is something that is cut off. The uterus does not have a decidual lining all the time, however. It develops from a more generic precursor in anticipation of implantation of an embryo. If pregnancy does not occur, it is shed.

Right away, we have questions: This seems wasteful; why not just keep that lining for the next time? And why form the decidua everywhere in the uterus—wouldn't it make more sense for it to be restricted to the spot where the embryo implants?

Another mystery relates to the evolutionary distribution of menstruation. In the previous chapter, I treated placental mammals as basically similar in terms of pregnancy. It is thus a bit surprising that very few others menstruate. Could this just be an artefact of human birth control? Nope, we can eliminate that one quickly: although women did historically experience fewer periods (because they spent most of their premenopause adult lives either pregnant or nursing), when female mice or rabbits are deprived of mates, they still don't menstruate. The basic reason for this is simple: in most mammals, it takes an embryo to kick-start the formation of the decidua. No mating, no embryo; no embryo, no decidual-ization. As a result, in most mammals the equivalent of our decidua only forms when it is needed. To call this tissue "decidua" is thus a misnomer, but biologists tend to see the world through a human lens.

Other menstruating mammals include our closest relatives, the primates, as one might expect. But they also include those that are much more distantly related: two distinct families of bats and the curious sengi (also known as the elephant shrew). It's hard to see what bats, apes/humans, and sengis have in common, other than being mammals. If them, why not all? Conversely, if so many other mammals don't menstruate, why do humans? When did this begin? The only way to answer these questions is by now familiar: we must combine the study of living species with the historical reconstruction of reproduction in our extinct ancestors, which is guided by evolutionary trees.

CONCEPTUAL THINKING

For all amniotes, including mammals, fertilization requires that sperm be released into the female reproductive tract by ejaculation, the muscular pumping that accompanies male orgasm. But how does the egg get from the ovary, where it develops, to the site of fertilization? That is, what regulates *ovulation*? The answer here depends greatly on the species. In many female mammals, there is a direct connection between mating and gamete release that mirrors what is seen in males. Mating can trigger ovulation by one of two different mechanisms.

Most commonly, as with rabbits and most carnivores, the mechanical stimulation of copulation is the trigger. More rarely, as for koalas and llamas, ovulation is instead a response to a soluble molecule transferred in the semen of the mate.

Regardless of the mechanism for ovulation induction, it makes intuitive sense. Even though the eggs of placental mammals are tiny compared to those of our oviparous ancestors, they are still in relatively short supply. Why release an egg if there is no sperm to fertilize it? Induced ovulation has a beautiful symmetry: in both sexes, the act of mating triggers gamete release. However, sometimes evolution has other ideas.

We primates have an ovulatory cycle that spontaneously (and in our case, cryptically) produces a fertilizable egg on a regular schedule, mating or no mating. In such a system, eggs can go unfertilized unless there is frequent mating. This predicts, then, that species with brief estrous periods and a short mating season will often be induced ovulators, and they are. Conversely, those with spontaneous ovulation, and especially those in which the female's fertile period is not signaled in an obvious way, will need to mate regularly. Consistent with this, chimps, bonobos, gorillas, and humans have converted sex from a brief moment once a year to a sort of regular pastime, in which the vast majority of copulations fail to produce offspring.

At this point, I cannot resist a wee tangent. It is fair to say that this recently expanded, highly social role of sexual activity underlies much of the joy and the anguish of being human. When we fall in love, the sexual aspect forms a big part of the excitement. When we lack (or lose) a partner, we feel palpable anxiety and loneliness, even when still surrounded by many friends. There is something unique about sexual intimacy and the relationships that go with it that we are wired to need, something that goes far beyond the immediate task of baby making. We long for it. We fight for it. We die for it. And we really, really like to sing songs about it. In some ways, we aren't really so different than other animals. After all, mountain sheep, birds, and dung beetles also fight and sing in the context of mating. But the fact that sex and the pursuit of relationships that offer sex is a year-round activity sets primates, and especially us humans, somewhat apart from our animal kin.

Back on topic, which form of ovulation came first—induced or spontaneous? If we place the many placental mammals, marsupials, and monotremes[1] with induced ovulation on a phylogenetic tree (figure 9.1), we see that this was likely the ancestral state for mammals.[2] The human condition, with spontaneous

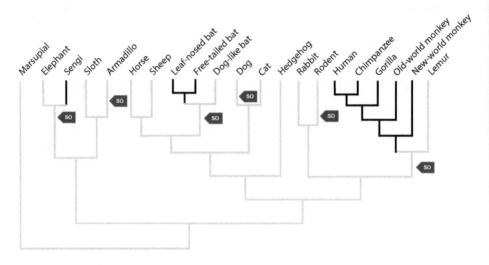

9.1 Multiple origins of menstruation in mammals. Species or groups of species with embryo-induced formation of the uterine decidua are shown in gray. In the ancestor of all mammals, this was accompanied by mating-induced ovulation. Switches to spontaneous ovulation (SO) are indicated by filled arrows. The subset of SO lineages that exhibit menstruation are indicated in black.

Source: Phylogenetic tree and reconstruction of reproductive evolution simplified from D. Emera, R. Romero, and G. Wagner, "The Evolution of Menstruation: A New Model for Genetic Assimilation," *Bioessays* 34, no. 1 (January 2012): 26–35, https://doi.org/10.1002/bies.201100099; M. Pavlicev and G. Wagner, "The Evolutionary Origin of Female Orgasm," *Journal of Experimental Zoology Part B: Molecular and Developmental Evolution* 326, no. 6 (September 2016): 326–337, https://doi.org/10.1002/jez.b.22690.

ovulation, arose in our lineage at the base of the primates. Similar shifts were made in parallel with several other groups, including rodents, artiodactyls (e.g. cows and whales), bats, dogs, and the Afrotheria group of mammals (hyrax, elephants, and elephant shrews).

All three origins of overt menstruation (in primates, bats, and the elephant shrews) occurred within this larger set of mammals with spontaneous ovulation.[3] That spontaneous ovulation is a prerequisite for menstruation makes sense. In the original system of induced ovulation, there is localized, embryo-induced decidualization of the uterine lining, and hence no menstruation is required. However, spontaneous ovulation is not sufficient to cause menstruation. This imperfect correlation exists because most species with spontaneous ovulation still require implantation of an early embryo to trigger uterine decidualization, as their ancestors did.

So we are still missing something: why would some species with spontaneous ovulation go further by turning their entire uterus into a potential landing pad for a tiny embryo and then bleed copiously if no embryo is present? Over the years, multiple ideas have been put forward to explain the value of menstruation. One that received a great deal of attention posited that menstruation was a sort of flushing mechanism, designed to remove pathogens that might be introduced during mating.[4] The idea's champion was Margie Profet, an unconventional scientist who created a media sensation and won a MacArthur Fellowship when she proposed it in 1993.

Despite the hype, the antipathogen idea is fraught with problems. To start with, why would such a germ-purging mechanism only occur with certainty in the complete absence of mating? University of Michigan anthropologist Beverly Strassmann produced a lengthy rebuttal in 1996, arguing that the predictions of the pathogen-removal hypothesis were generally not met. For example, the intensity of menstrual flow in primates is not at all correlated with the number of mates, as would be expected if it were an adaptive hygienic response. Over time, the antipathogen hypothesis has fallen out of favor.

If not there to rid us of germs, then why do some animals with spontaneous ovulation endure monthly bleeding? Is it directly beneficial, or a consequence of something else? Strassmann proposed that menstruation occurs when the degree of investment prior to implantation become so great as to be a major metabolic burden should pregnancy not occur. In other words, retention is ultimately more costly than shedding.[5] In 2011, Yale biologists Deena Emera and Günter Wagner further argued that such a costly-to-maintain endometrium arises when there is extensive spontaneous decidualization (SD) of the entire uterus. In such cases, it becomes impossible to resorb the large amount of blood-rich tissue completely. The only alternative is to shed the excess.

Because not all spontaneous ovulators have SD, why evolve it if it necessitates menstruation? Emera and Wagner suggest several ideas. One is that SD is a response to a tendency of the embryos of some species to produce an especially invasive placenta. Invasiveness maximizes contact between fetus and maternal nourishment, allowing vigorous fetal growth. But it can also harm the mother's longer-term fertility and health if left unchecked. By making a thick inner uterine layer in advance of the embryo's arrival, this fetal-maternal conflict is resolved. The demands of a greedy conceptus are met, while deeper tissues are protected, allowing for both a successful pregnancy and for future pregnancies.

It has also been proposed that SD might allow for a female to judge the health of newly implanted embryos. Intriguing experiments conducted by a Dutch team on cultures of early human embryos with uterine stromal cells suggest a quality control check is made at the time of implantation.[6] While normally developing embryos enjoyed a stable local environment, abnormal embryos triggered a local cessation of uterine growth factors required for ongoing fetal growth. The ability to recognize abnormal embryos was only seen in uterine cells that had undergone decidualization, however. This suggests that preimplantation SD enables rapid screening of embryos. Whether due to parent-offspring conflict or embryonic quality control, the absence of pregnancy necessitates elimination of the lining, likely for the reason Strassmann noted.

EGGS, ORGASMS, AND ANATOMY

In our species, the female orgasm might seem a curious thing. Sure, it is nice, yet it is not required for making babies and unfortunately often fails to occur at all in heterosexual intercourse. So why does it exist? It could be a hedonistic induce-ment to mate, even if the payoff is ultimately often lacking. But remember, we are but one of many mammals—are we representative? A clue "comes" from stud-ies on women volunteers, who agreed to have their blood monitored before and after orgasm. This revealed a surge of the hormones prolactin and luteinizing hormone (LH).[7] Across vertebrates, spikes in these same hormones are also asso-ciated with ovulation and are also known to be induced by mating (and orgasm) in other mammals.[8] The signs all point to the human female orgasm as a vestige of an earlier time when it was actually needed to release eggs for fertilization. This original state of things had a satisfying symmetry: male and female climax served the same function of releasing gametes.

We can go further to connect orgasms to menstruation. Orgasm in both sexes relies on stimulation of the developmentally equivalent penis and clitoris (see chapter 2). Mihaela Pavlicev (now at the University of Austria) has pointed out that the anatomy of the female genitalia is shaped in a predictable way by the mode of ovulation. In most species with induced ovulation, such as the gray short-tailed opossum, the hedgehog, and the woodchuck, the clitoris is posi-tioned inside the vagina. This maximizes the stimulation it receives during mat-ing, ensuring that their equivalent of orgasm occurs and the hormonal responses required for ovulation that go with it. In contrast, species with spontaneous ovulation usually (but not always) have a clitoris that is rotated out of the vagina. This includes most species that menstruate. Despite its different roles in ovulation, Pavlicev's research indicates it is a reflex with a common neurophysiological basis across mammals.[9] "The same chemicals that suppress orgasm should work in both sexes. Prozac does this in humans, and in rabbits it blocks ovulation."

The above correlation between clitoral position and mode of ovulation presents a mystery: if orgasm is pleasurable, then it can serve as an inducement to mate, even if it's no longer needed for ovulation. Leaving the clitoris in the vagina, where that sensitive bit is in the most advantageous place for feeling, seems like it should always be preferable to shifting it outside, where so much more uncertainty occurs. This delicate mystery remains unresolved, but Pavlicev suggests one possibility. "Most species don't mate outside of getting pregnant. We conceal our ovulation and mate all the time. Mice wouldn't even mate if they are not in estrous. Perhaps the logic is that having induced ovulation means you may want to not ovulate with every copulation." In other words, as the frequency of mating increased in our ancestors, there may have actually been a problem with inducing too many ovulations. The shift in clitoral position would limit them to a manageable number. Orgasm and the timing of ovulation would eventually be separated completely, at which point sex is freer to take on a social role. At the extreme end, a study of free-ranging Japanese macaques found that females mount other females many times per hour, further stimulating themselves with their tails as they do.[10] This wouldn't be possible were ovulation still tied to orgasm.

With orgasm no longer serving its historical function as an ovulation trigger, why keep it? One possibility is the developmental and physiological connection to males. Male and female bodies have corresponding parts that go back to embryonic stages (see chapter 2). Males have useless nipples because there is no reason to suppress their formation. Similarly, male orgasm is essential for fertility and is connected to penile stimulation, so perhaps the clitoris retains orgasmic potential in species that lack induced ovulation only because it absolutely must be there in males. In this scenario, because orgasm causes no harm in females, it is a sort of sex-specific pleasure appendix.[11]

On the other hand, human females may retain the ability to have orgasms because they indirectly contribute to successful conception and child-rearing. I favor this hypothesis. The prospect of an orgasm is part of the general sex drive and must incentivize mating to some extent. Mating is unarguably essential for pregnancy, at least in a premodern world. In addition, birth is only the beginning in our species—human babies take years of investment to rear. The longer copulation or alternative techniques required for human female orgasm may reinforce pair bonding and encourage mates to try repeatedly to "get it right." In this way, sexual activity goes from a furtive moment to a year-round activity. For a species like ours, in which (a) nobody knows when conception can occur and (b) the rearing of the offspring requires parental cooperation for over a decade, this may be crucial.

"BLOOD COMING OUT OF HER—WHEREVER"

The August 2022 issue of *National Geographic* featured a poignant photo essay by photographer Sabiha Çimen. Her subjects were Turkish teen girls in a Quran-focused school, similar to one she had attended herself. The photos juxtaposed the girls' conservative dress and solemn religious study with aspects of twenty-first-century life, capturing their imagination, humor, and yearning. One of the seemingly more conventional photos depicted a student, Reyvan, with a platter of tomatoes she had gathered to help the school's cooks. But the caption jumped out to me in the context of this book: she was on tomato duty because touching the holy Quran during her period was forbidden.

The view that one's period makes one impure is widely held in Islam, although scholars disagree about the implications for religious practice.[12] Orthodox Jewish women face similar constraints. In Leviticus (15:19), the Lord instructs Moses and Aaron that "when a woman has a flow of blood from her body, she shall be in a state of menstrual uncleanness for seven days. Anyone who touches her shall be unclean until evening." Widening our view to the world at large, menstrual taboos and lack of hygiene products impose a huge burden on women nearly everywhere.[13]

Given that a typical woman will need period products five days per cycle, hits menarche (i.e., has her first period) at age thirteen, and reaches menopause at fifty, she will need period products for 2,470 days of her life if she doesn't have children. In 2024, a pad or tampon costs about 15 cents in the United States. How many are needed per day varies, but if we assume four then a woman will spend about $1,500 on pads or tampons over her life.

That number can be seen in two ways. First, it is small enough that developed countries could easily make them available to their residents for free. Scotland and New Zealand already do. Until that happens consistently here in the United States, three Girl Scouts in Vienna, Virginia, have taken matters into their own hands. In the spirit of the Little Free Libraries and walk-up food pantries, Isabel Buescher, Ariyanna Ghala, and Ramsey Warner gathered donations to build and stock a "period pantry" outside a local church.[14]

The other way to see the cost of period products is that, for poor women, it is simply far too high. Either it represents a nontrivial female-specific tax, or products cannot be obtained at all. Indeed, the United States Agency for International Development (USAID) reports that 500 million girls and women around the world lack the resources to manage their periods.[15] Developed countries can therefore make a big impact in the lives of women by using development aid to address the problem. The United States recently began working with Nepal, Madagascar, and other nations in this area, but clearly there is much to be done.

Donald J. Trump, America's forty-fifth president, was also famously disturbed and threatened by the prospect of a menstruating woman.[16] While many may not enjoy their period, you can by now see that it is ancient process, one inextricably connected to many other aspects of our reproductive biology and its history. We cannot eliminate all discomfort of a woman's period, but we can work to destigmatize it, and to combine sympathy and a bit of simple, inexpensive technology to eliminate the barriers it too often imposes. Rather than freak out or avoid talking about it, let's lean into it. Heck, we could even celebrate it. This actually occurs in some societies,[17] and ours could be one, too.

THE MENOPAUSE MYSTERY

In most vertebrates, including birds and most mammals, females can produce eggs until near the end of their life. In humans, however, the number of egg-producing follicles in an ovary is finite, and eventually they run out. After decades of monthly ovulation, the cycle becomes irregular and then ceases altogether. In the United States, the average age at menopause (defined as one year after the last period) is fifty-one. This marks the end of a woman's reproductive life in the baby-making sense (although caregiving may continue for much longer). Menopause frees her from menstruation, which might seem a perk, but introduces its own challenges. Egg follicles, in particular the granulosa cells around the egg, are the source of most estrogen in a woman's body. As a result, estrogen levels plummet during the transition to menopause, which can take years and is often accompanied by hot flashes, difficulty sleeping, and mood swings. Once definitive menopause is reached, most women experience changes in their skin and in their body's shape, and they begin to lose bone density (leading to osteoporosis). Some women also develop heart disease. In short, menopause is not fun.

Given that many women live for four decades after menopause, and given that men do not generally experience a similar loss of reproductive capacity, why does this happen? Is menopause an adaptation shaped by natural selection or an incoherent consequence of some other dynamic? Answers to these questions have been put forward by many and fall into two broad camps.[18] On the one hand, menopause could be selected directly for its ability to increase reproductive success, in the wholistic sense of rearing offspring to adulthood. The inability to give birth while there is still substantial vitality left ensures that the last offspring that are born to a mother, which need over fifteen years of care or more, are seen through to maturity. This is a better bet than having babies shortly before death, babies that would be likely to die as well. This benefit could also be extended to grandmothering—helping one's children raise their own children is a norm worldwide.[19]

The alternative to the above adaptive model posits that there is no real adaptive benefit to menopause and that it evolved instead because of the rapid changes in in human life history relative to our primate ancestors. Humans live much longer than any other primate, and it is possible that the increased egg supply required to extend fertility to near the end of life was not able to catch up. Evolutionary theory predicts that reproduction late in life is less valuable to a lineage of organisms than is reproduction early in life because early births can produce early grandchildren, creating a relatively short generation time that can maximize population growth.[20] Late offspring are the converse, and thus natural selection is thought to favor early reproduction strongly and be relatively impotent to shape it late in life. The implication for humans would be that menopause exists because the adaptive benefit to preventing it is so weak that it cannot budge the status quo. That men do not have a similar decline in sperm production could be explained by the difference between male and female gonads: testes are sperm factories with indefinitely renewable stem cells.

In considering the two alternatives above, I favor the adaptationist side. For starters, the age of menopause is rather tightly clustered, with essentially nobody able to become pregnant past their mid-fifties. If this trait were not under selection, like, for example, the graying of one's hair, we might expect to see a sixty- or seventy-year-old give birth occasionally. We don't. Second, the idea that evolution would be unable to adjust oocyte number in response to our longer life span is hard to believe. Chimpanzee females also stop ovulating at about fifty, but this is also their maximal normal life span. Only a modest change in oocyte number would be required to nudge humans into a similar match between fertility and life span, but it didn't happen. Finally, people are unique, even among primates, in the great length of time required to rear our young. While classic theory predicts that having a baby at age seventy has a small benefit, it seems intuitive that it is actually a net negative because of the harm it would cause to previous immature offspring. This would strongly select for a block to late pregnancies, and this matches what we see.

CHAPTER 10

Scents and Sensibility

Synopsis

Besides the sense of smell we recognize, most mammals have a vomeronasal organ (VNO), a separate "sex nose" that detects pheromones. As humans and our closest primate relatives evolved overt sex differences that made pheromones less important, the VNO was lost. The resulting visual orientation permeates modern life.

———————————————

Like a lot of other families facing COVID-19 lockdowns in the spring of 2020, we got ourselves a quarantine puppy. Walking our fluffy Shiba Inu, Yuzu, around the neighborhood, it is clear she takes in a lot of information we humans cannot sense. She carefully sniffs spots where other dogs have peed, savoring their nuances like a snooty wine connoisseur. And then there is the mutual butt sniffing when we meet another dog. From this, she gains a great deal of information, including the sex, reproductive status, and specific identity of the other dog.

Dogs' ability to sense information about their fellows through smell relies on a dedicated pheromone-detecting part of their olfactory system. Neurons specialized for sensing pheromones are localized to the vomeronasal organ (VNO). The VNO is located in the palate, closer to the snout than the main olfactory epithelium used for nonsocial smells (like food). It sends its information to a distinct part of the brain, the accessory olfactory bulb (figure 10.1). In a very real sense, dogs (and many other mammals) have a dedicated sex nose. The VNO seems to have arisen shortly after vertebrates moved onto land because it is also found in

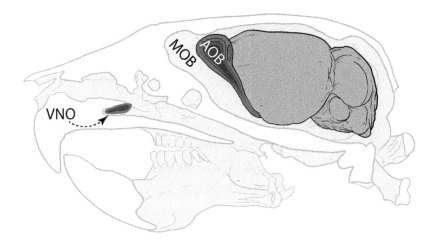

10.1 The pheromone-sensing olfactory system of mammals. Pheromones entering the nose are sensed by the vomeronasal organ (VNO), which sends this information to the accessory olfactory bulb (AOB), a region of the brain next to the main olfactory bulb (MOB).
Source: Nicholas Bezio.

amphibians and most amniotes. This allowed them to exploit the new sensory opportunities provided by life in the air.

The function of the VNO has been best characterized in lab mice. It is actually rather challenging to tell the difference between a male and female mouse. They are about the same size and color, and their genitalia are not easy to inspect (even if you wanted to). However, thanks to their VNO, the mice themselves have no trouble recognizing each other's sex. Even in the dark, a male mouse will quickly determine whether another mouse is a female of reproductive age and will attempt to mate with her if she is. If the other mouse is a male, however, a brief fight often breaks out. The VNO appears to be crucial for much of this selectivity. A landmark study by the lab of Harvard's Catherine Dulac examined mutant male mice lacking a protein called TRPC2, which is required for the VNO's sensory activity.[1] These males no longer acted aggressively toward other males and attempted to mate with both males and females indiscriminately. A later study found that females bearing the same mutation also exhibit male-like behavior toward both males and other females, such as intense butt sniffing and mounting.[2] Surgical removal of the VNO produces similar results. This all indicates mice need their VNO to tell the sexes apart and to behave accordingly.

If you have noticed that butt-sniffing is not a routine part of human social interaction, that is because we wouldn't get a lot from it. While most new-world monkeys retain a VNO, none of the Catarrhini (the group that includes old-world monkeys, apes, and humans) have one anymore.[3] Embryonic development provides another, independent confirmation that our ancestors once had a VNO. Although it starts to develop in human embryos (figure 10.2), the final structure in adults is variably present, and lacks sensory neurons.[4]

We aren't the only mammals to have lost their sex nose. For example, whales, dolphins, and manatees also lack a VNO. Smell isn't much use under water, so this makes sense. Why did our ancestors give up theirs? The answer is not as clear as it is for the aquatic mammals, but several lines of evidence suggest it is related to a shift of our social interactions away from smell and toward vision (a topic explored more fully in the next chapter).

In our distant primate relatives, males and females are often not very different from each other. For example, male and female flying lemurs, lorises, and true lemurs are essentially identical in size. They also lack obvious differences in pigmentation or other ornaments.[5] In contrast, the males of nearly all old-world

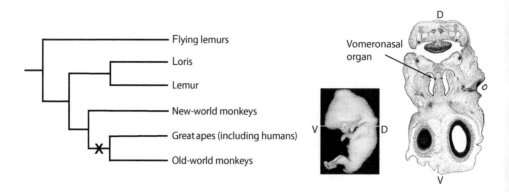

10.2 Loss of a functional vomeronasal organ (VNO) in the great apes. At left, a simplified evolutionary tree of primate relationships, with the loss of the VNO indicated. At right, the paired pits of the human embryonic VNO of Carnegie stage 22 (eight weeks), as seen in a section through the palate. The section's location is indicated in the schematic at lower left. Only the ventral (V) portion of the section is reproduced here. The VNO remnants found in many adult humans are nonfunctional.

Source: Embryo images from the Boyd collection embryo H983 (courtesy of the Endowment for Human Development).

monkeys and apes are substantially larger than the females. In addition, there are often male-specific ornaments in apes and old-world monkeys. Perhaps the most impressive is the red facial pigmentation of the mandrill, which is so bright it looks as if it had been painted on. Male and female voices are also different in pitch and timbre. It would seem our ancestors long ago became able to tell the sex of another individual without smell and thus had no use for the VNO.

To get a sense of just how attuned we are to recognizing sex differences visually, consider the 1977 study of Lynn Kozlowski and James Cutting of Wesleyan University.[6] They filmed male and female student volunteers as they walked toward a television camera in a way that obscured most information about their body shape. The volunteer walkers wore dark clothes with highly reflective tape around their joints. Then they were illuminated by a very bright light set up near the camera. The camera was fed into a monitor whose contrast was turned up to the point where only the reflective spots were visible against a black background. The result was a completely abstracted view of a walking form, with only dots on a black background. The researchers then showed other subjects the videos they had made, either as static images or clips of moving dots.

When Kozlowski and Cutting asked their viewer subjects what static images of the dots represented, few guessed they were people. Once told what the dots represented, they could not determine the sex any better than random from stills. However, once an animated "dynamic display" was played, they found people became quite good at distinguishing the sex of the walker. Male walkers were somewhat easier to recognize by both sexes of viewers, and the ability to distinguish sex depended crucially on the swinging of the arms: movies taken with the walkers' hand in their pockets were not informative. The researchers concluded, "It is clear that a dynamic display of point-lights is sufficient for recognizing the sex of a walker."

The above study emphasizes how attuned we are to subtle visual indicators of sex. With full illumination and the overall shape of the body visible, humans can often distinguish adult males and females even at a fairly great distance and especially if they are moving. This ability likely stems from the different skeletal proportions of adult male and female humans. However, it also creates social feedback related to gender norms. A quick internet search shows that the femininity or masculinity of one's gait is a preoccupation of both cis- and transgender people. How we walk is therefore shaped by both our anatomy and our own awareness of how our gait is perceived by others (more on this, as it applies to the brain, in chapter 12). After reading Kozlowski and Cutting's study, I started playing a game when I walk around outside: I pick a far-off person walking toward me, try to guess their sex (or gender), and then see whether my impression remains as they pass me. Sometimes I'm clearly wrong, and sometimes the "right" answer is not obvious even up close. Most of the time I'm right, though. Try it yourself and see how it goes.

DO OUR NOSES MAKE US QUEER?

When the *TRPC2* mutant mice were first described, much was made of their greatly increased tendency to engage in same-sex sexual behavior (SSSB). Like the other old-world monkeys and apes that lost a functional VNO, we lack the *TRPC2* gene entirely. It's probably not news to you that human SSSB is also widespread (see chapter 13). It was therefore only a matter of time before a connection between the loss of *TRPC2* and human sexuality would be proposed, and in 2019 Daniel Pfau and his colleagues at Michigan State University put it out there. Noting that *TRPC2* loss and frequent SSSB appear at the same point on the primate evolutionary tree, they proposed that the former enables the latter. They also note that *TRPC2* mutant mice can still be trained to recognize the sex of other mice, suggesting that SSSB in these mutants is not solely about misdiagnosis.

Having corresponded some about the evolution of human sexuality with the editor of the journal that published the conjecture, I was asked to comment on Pfau et al.'s hypothesis. I had to admit that the correlation of VNO and *TRPC2* loss with frequent SSSB is noteworthy and that it tells us that something did indeed change. But did it predispose humans to SSSB, and especially to the exclusive preference for same-sex partners? That is, did the loss of VNO function allow for some of us to be gay? I concluded that it probably did not, for several reasons.

For starters, another thing that evolves at about the same time as *TRPC2* and the VNO are lost is exaggerated nonolfactory sexual dimorphism. Size, body and facial shape, gait, hair and pigment differences, and vocal pitch all contribute. We apes do not need a VNO to distinguish the sex of our fellows reliably, and thus one of its key functions was moot. It may have only been a matter of time before the VNO, and eventually the *TRPC2* gene that is crucial to its function, were lost. We cannot know the precise order of events, but we cannot reject that *TRPC2* loss was a consequence of these other changes rather than a cause.

There is still the question of why SSSB becomes especially frequent in old-world monkeys and apes, and how that might relate to human homosexuality. *TRPC2* may have inhibited SSSB independently of sex discrimination per se, so that remains possible in my mind. However, there are strong indications that SSSB in humans and nonhuman primates are not the same thing. SSSB in nonhuman primates is both hedonistic and a means of enforcing dominance hierarchies. It exists alongside procreative sexual activity (i.e., actual heterosexual mating).

In our own species, hedonic SSSB is found even in people with a primarily heterosexual orientation, especially during adolescence.[7] However, from the pioneering work of Alfred Kinsey and coworkers and others,[8] we know that a large percentage of adult men and women have a strong orientation toward *exclusive* homosexuality. For men, exclusive homosexuality is more frequent than bisexuality, it develops early in life, and it remains in the face of strong discouragement.[9] This all suggests that gay people are not manifesting the same behavior as nonhuman primates who engage in SSSB alongside reproductive sex. It is something new, and why it may have evolved is the topic of chapter 13.

VISIBLY SEXY

Show a dog or a mouse photographs of potential mates or competitors, and they would show no interest whatsoever. Presumably this is because they don't *smell* like dogs or mice. The shift to more visual sex recognition from the olfaction-dominated mode of our recent ancestors could be seen as just another of the many interesting twists and turns on the road to us. Its impact on modern society, however, suggests it is much more than that. What would the world be like if male and female bodies were indistinguishable outside periods of pregnancy and nursing? An androgynous world in which discreet inquiries must be made to know a coworker's sex would surely feel very different.

In our actual world, we humans have a habit of checking each other out, even in representations that are clearly not real. Michelangelo's David? Kinda hot, even though he is literally a rock. This extends even to two-dimensional forms. When you walk by a newsstand or wait in the grocery store checkout line, notice what is mostly on the cover of the magazines offered for sale: sexy people. These magazines sell because we can't help but look. How we move, the sound of our voice, and our general reputation matter, but I will argue that the visual reigns supreme, inasmuch as it is sufficient to attract interest. The proliferation of porn, initially in print and now on a massive scale on the internet, is spectacular proof that we can be aroused with only sights or sounds.

As countless internet influencers are happy to explain, there are many tricks to enhance one's looks, whether in person or in a still image. The tools seem to fall into two basic categories. One draws attention to visibly sexually dimorphic features of the body itself (whose origins are the topic of the next chapter). The other category is bling:[10] clothing, makeup, and jewelry can directly exploit the human visual bias independently of (or synergistically with) the bodies they adorn (figure 10.3). Bling can also communicate things other than narrow-sense sexual attractiveness, such as power, wealth, and creativity. These traits cannot be completely divorced from sex, however, because both resources

10.3 Adornments as attractors in contemporary society. Beyond highlighting female- or male-typical body traits, clothing, as in this ensemble from designer Marc Jacobs (*left*), and jewelry by Tucson artisan Sam Patania (*right*), draw attention in other ways.
Source: (*Left*) Randy Brooke, used with permission of Getty Images; (*right*) Sam Patania.

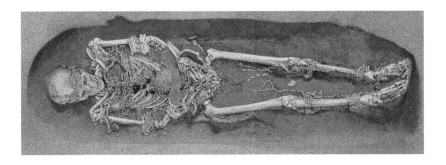

10.4 The Sunghir 1 burial from Paleolithic Russia.

Source: E. Trinkaus and A. P. Buzhilova, "The Death and Burial of Sunghir 1," *International Journal of Osteo-archaeology* 22 (2012): 655–666, https://doi.org/10.1002/oa.1227.

and resourcefulness are also part of mate choice. This tendency to draw attention to one's form, both its anatomical parts and through adornment with clothing and jewelry, is extremely ancient. For example, a Paleolithic man buried roughly 25,000 years ago in what is now northern Russia was covered head to toe in elaborate necklaces (figure 10.4).[11]

I think it's fair to say that women are the most frequent subjects of visual celebration. Fashion in the broad sense is a disproportionately female interest (or, some may say, obligation). The voice is also part of the feminine allure. A beautiful and skilled female voice is the core product of pop music, for example. From my perspective as a zoologist, however, this is actually really curious. In birds, for example, the male is always the more splendid specimen, both visually and in terms of singing. The same pattern holds for insects, fish, and many other animals. That sex-typical ornaments are somewhat reversed in humans is thus noteworthy, and this phenomenon will be addressed in detail in chapter 11. In short, a likely explanation is that once pair bonding became a norm, females evolved features that both attract a mate and keep him interested long after an offspring's conception.

The associations between our body's shape, how we move and speak, our choices of clothing and accessories, and our internal sense of our gender all contribute to what others perceive as our gender expression. Gender identity will be discussed more fully in chapter 14.

CHAPTER 11

The Male Gaze

Synopsis

The males of many animals are highly ornamented, and this makes them attractive to females. This results from the evolutionary phenomenon, first formalized by Charles Darwin, of sexual selection. In humans, both anatomy and culture suggest that women also began to experience new forms of sexual selection after we diverged from our closest living relatives. This unusual dynamic contributes to (but cannot excuse) the persistent objectification of women's bodies.

———◆———

I was near the end of sixth grade when it started:

> ME: "Hello."
> THEM: "Doug?"
> ME: "Uh, no, sorry, it's Eric. Let me go get him."
> THEM: "Oh, Eric! You sound just like your dad. Golly. How old are you now?"

Far more disturbing than people mistaking me for my dad on the phone, weird hairs began sprouting out of places where there were none before. The common showers we were obliged to use after physical education class confirmed I wasn't alone, and they got more and more awkward. We were not all on the same schedule, and one could be teased equally for being among the first to become hairy and for remaining boyish longer than most. Through the beige cinder block walls that divided our locker rooms, the girls were similarly transforming into women, a process which began a few months earlier than for us.

When we hit puberty, our sexual selves become visibly, audibly, and olfactorily obvious. Our parents and everyone with whom we interact notice, and they have a hard time not pointing it out, even when they really shouldn't. Take, for example, a kind of comment often made to a father by his peer male friend, "Wow, Jennie sure is growing up!" (usually followed by a nervous chuckle). This generally is taken by the father to mean "Wow, your daughter used to look like a little girl, but now she is curvy and I find myself involuntarily evaluating her as a possible sexual partner. I admit that's creepy and I feel guilty about it, but maybe in some odd way you are reassured to know your daughter is becoming an attractive woman? Sorry, I should stop talking now . . ." Beyond these outward appearances, our adolescent brains become occupied with crushes and fantasy, and sex pervades our interactions with our friends. Simultaneously exciting and bewildering, it changes our lives forever.

Many of the changes we experience at puberty relate directly to reproduction. For example, little boys do not produce mature sperm or seminal fluid. In adolescence, the testicles and prostate grow and begin to function, alongside a growing inclination to try them out. Similarly, little girls have ovaries, but no mature eggs are released for fertilization, nor can the prepuberty uterus support pregnancy. The onset of menstruation marks the start of the monthly fertility cycle, which develops alongside behavioral interest in sexual activity that parallels that of boys (see chapter 9 for more on that). However, if these fertility-enabling changes were the only ones we experienced at puberty, it would actually be rather difficult to determine an adult human's sex if their genitals were covered (at least outside pregnancy). What makes it fairly easy is a set of other, *secondary* traits. In response to sex hormones produced by the testes and ovaries, the binary differences in reproductive equipment are augmented by differences in the pitch of the voice, the shape of the face and shoulders, whether that face is covered with a beard, relative muscle mass, and localized fat deposits. These traits vary somewhat from person to person, but collectively they signal the maleness of a man, and the femaleness of a woman, beyond what is needed to make babies.

Although it seems normal to have secondary sexual traits, many animals lack them. Why do we have them? How old are they, in an evolutionary sense? What can they tell us about how humans have paired up for reproduction (i.e., our mating system)? What do they have to do with attractiveness to others, and how has this shaped our anatomy, behavior, and present-day culture? The aim of this chapter is to provide provisional answers to these questions, drawing on research from physical and cultural anthropology, evolutionary biology, and psychology. The overall conclusion is that some human secondary sex differences have deep roots in the primates, yet we also differ from our closest relatives. These recently evolved differences are especially evident regarding women.

SEXUAL SELECTION: A PRIMER

Charles Darwin's *On the Origin of Species* was important for two reasons. First, it laid out an impressive collection of evidence for the existence of evolution, what he termed "descent with modification." He was not the first to propose the existence of evolution, but his arguments were far more persuasive than those of his predecessors. Second, he was the first to describe a mechanism for evolution: the process of natural selection. Organisms that are well suited to their environment and way of living, facing as they do a continual "struggle for existence," will tend to leave more offspring than those that are not. If they, too, are like their parents, they come to dominate. This is the plain vanilla sort of selection that explains the remarkable fit we see between an animal's mouth, appendages, and sensory organs and how they make their living. Important for us here is that Darwin also pondered traits that exist in only one of two sexes, traits that are not helpful in the struggle for existence. The peacock's splendid tail is a classic example: its growth takes six months and adds roughly 10 percent to his basic metabolic needs, yet it has no daily function.[1] In a second classic volume, *The Descent of Man and Selection in Relation to Sex*, Darwin proposed that such traits result from the process of *sexual selection* acting within a species. The basic idea is that if one sex has a preference for a particular trait (tail feather displays, songs, etc.), then individuals that produce that trait will have more mating opportunities than those that don't. Over time, this leads to the spread of the trait, as long as its associated costs are not so severe as to outweigh their benefit.

Here is how sexual selection can lead to the spread of traits that are otherwise costly: Suppose growing a splendid tail reduces a peacock's adult life span by 10 percent. Assuming paternity is similar every year of life, this would be expected to reduce his offspring by the same 10 percent. A bad deal, right? But what if eliminating the tail to regain that 10 percent in longevity reduced his opportunity to mate by 20 percent? If we think of the no-tail life span combined with the with-tail mating success as the optimal combination, then we see that having the tail gets the peacock to 90 percent of that, while eliminating the tail can only get the male to 80 percent. As a result, tail-bearing males leave more offspring, and the trait spreads despite its harmful effect on life span. The existence of sexual selection creates a more inclusive formulation of natural selection, in which *any* traits that increase the overall chance of successful reproduction will be selected, even if they are not required for fertility per se, and even if they begin to introduce costs in nonreproductive aspects of living.

The peacock is a case of males being selected for their attractiveness by "choosy" females. Female choice is the most commonly studied (and commonly taught) sort of sexual selection. It is especially powerful if the selected male gets to mate with multiple or even all females in the local social group, to the exclusion of other males who are not selected. This kind of system is referred to by biologists as *polygyny*. It is found in diverse animal groups, from dung beetles to deer, from salmon to gorillas.

Competitions between males for access to females are not always just beauty contests. Antlers, horns, canine teeth, and large size are used either to threaten or to engage in direct combat with rivals. Having intimidated, injured, or driven away the competitors, the victor gives the females only one choice. Yet despite the very different dynamic, the outcome is essentially the same as with female choice: the oversupply of males exerts strong sexual selection on male traits.

In the choosy female paradigm, what (if anything) can the female achieve by choosing? There have been a number of explanations. One is that the male trait may directly signal his health and vigor—the so-called good genes hypothesis. Another is that, because males with showy traits will be attractive to most or all females now and in the future, a mother-to-be will choose such males because her sons will be similarly attractive. This is referred to as the sexy son hypothesis. Perhaps the wildest idea was put forth by Amotz Zahavi in 1977. Zahavi suggested an extreme form of the good genes model: some traits may be so conspicuous that their very existence implies an unavoidable cost to the male, but one that the most vigorous specimens could easily pay. As my grad school evolution professor, Curtis Lively, explained it in an especially memorable lecture, this could work by effectively communicating "Hey, I'm so gnarly I can make it out here with these big honkin' horns on my head!" What has come to be known as

Zahavi's handicap principle has generated controversy among evolutionists, but it may hold in some situations.[2]

Sexual selection acting on male traits, whether driven by female choice or male-male competition, is expected as a consequence of anisogamy (see chapter 3). Females invest much more in babies than do males (either by making a big egg or through pregnancy), and so they have a much smaller maximum number of offspring. Males, however, have the potential to monopolize a large group of mates, creating a sort of genetic sweepstakes that selects for males that can pull it off. A recent meta-analysis of sexual selection by Tim Janicke and his colleagues demonstrates that males experience stronger sexual selection than females across the diversity of animals, a trend they have dubbed the "sex role syndrome."[3] This confirms that Darwin's basic intuition was correct. However, there is also tremendous variation in how sex develops, and there are multiple exceptions to the "males fight or charm, females choose" and "males are fancy, females are inconspicuous" paradigms.[4] These exceptions are instructive, and our own species is one of them.

MALE CHOICE AND SEXUAL SELECTION ON FEMALES

Just as female choice and male-male competition can select for male traits, there are some species in which male choice and female-female competition lead to sexual selection on female traits. But if a male can mate with unlimited females, why would males ever be choosy? The clearest cases relate to situations where roles are reversed: in some species, males cannot have many mates. For example, male pipefish (who are defined as male by being the sperm donor) place fertilized eggs into their abdominal brood pouches and essentially become pregnant. This is something like a buck kangaroo carrying the joey, except there are roughly a hundred of them in there. Because female pipefish lack their own brood pouches, males become a limiting commodity, and this leads to the males being choosy in two ways. First, they are more likely to mate with larger females, and after mating they preferentially nourish the eggs of these larger mates.[5] This should select for larger females, and indeed females are substantially bigger than the males. Male pipefish also prefer females with pretty stripes on their sides.[6]

A similar dynamic is seen in comb-crested jacanas, lovely Australian shorebirds with jaunty red ornaments on their heads. As with other jacana species, females lay clutches of eggs for their mates to incubate before moving on to another male and another nest. Male jacanas who leave the nest to seek another mate risk losing the offspring they already have sired and thus tend to be choosy about their mates. This leads to competition between, and thus sexual selection of, females.[7] As with the pipefish, males prefer females of larger size.

Besides the species with choosy males, there is also a female counterpart of within-sex competition. With rare exceptions stemming from an acute shortage of males, this competition is less about fighting over access to mates and more about fighting for limited resources needed to reproduce successfully.[8] For example, spectacled parrotlets only nest in cavities in trees. It is imperative to find a suitable cavity; they can be hard to find, and fights between females over the best spots can occur.

When females are attempting to attract a male or are competing with other females, are they aided by the same sorts of ornaments that males use to signal their overall size and health? There is some indication they are. Striking ornaments (some of which could also be weapons) exist in many female birds, fish, and insects. In most cases, however, these are not female-specific, as one might otherwise expect. For example, both sexes of the jacanas noted above sport their namesake red head crest. Is this because it benefits females, but males develop it anyway because no harm is done to them by possessing it? Or might it be present in males for species recognition? The possible roles of sexual selection of ornaments shared by both sexes are not clear.

Are there cases of ornaments that only females have, something like a female version of the peacock's tail? This is a very rare situation, but there are examples.[9] Those not associated with sexual role reversal (as with pipefish) are especially rare, but they also exist. In another parrot, *Eclectus roratus*, males are green and females are bright red and blue. Males spend all day among the green foliage, whereas females live most of the time inside nest cavities. These cavities are in short supply, and females guard them carefully. Here, it seems males have lost red pigment in response to natural selection to blend in rather than strong sexual selection to generate new colors in females.[10] The females may retain the more conspicuous colors to aid in female-female competition for rare nesting sites.[11]

Another class of female-specific ornaments is a transient signal of reproductive phase. Female baboons develop outrageously swollen vaginal openings as they near ovulation and are more receptive to mating at this time. This trait ensures the window of fertility is not missed by encouraging males to mate. It could also be regarded as a form of female-female competition because females unable to produce the swelling would be less likely to become pregnant. It could also be seen as a form of male choice because males who are attracted to the swellings are more likely to become fathers than those who are indifferent.

In Lake Eyre dragon lizards of Australia, female-specific fertility signaling comes with an odd twist: the orange belly pigment that signals fertility persists well past its end so that inseminated females continue to be approached by would-be mates. The females resist these advances, often by flipping on their backs and showing their orange bellies.[12] In another lizard, *Microlphus occipitalis*, a red throat is only present in the postfertile period. In both lizards, these ornaments have been proposed to serve as the opposite of an attractant—a sort of "don't try to mate with me!" signal.[13]

SEXUAL SELECTION AND PEOPLE

Sexually selected traits first appear in adolescence; are useless in the general sense; and yet are a subject of fascination by others, especially the opposite sex. It's not too hard to see that we humans have such traits ourselves. Some seem to follow the conventional paradigm of female choice and male-male competition. Facial and body hair and lowered vocal pitch are indicators of adulthood in men, as is a squaring of the face and growth of the Adam's apple. These features signal maturity to both prospective mates and to other men, and they are triggered by a surge in testosterone associated with development of the testes and the onset of fertility.

Another change in males at puberty is a substantial increase in skeletal muscle. A muscular (but not *too* muscular) body is directly attractive to females from countries as diverse as China, New Zealand, the United States, and Cameroon.[14] It also mirrors sex differences in strength and size seen in a more extreme form in primates with prominent male-male competition, such as gorillas. The sex difference in strength in our own species (discussed in detail in the next chapter) is thus both a result of female preference and a sign that male-male competition (arm wrestling? something rather worse?) has been part of our lineage for a long time.

In addition to garden-variety sexual selection on men, however, there is something else going on in *Homo sapiens*. Take breasts, for example. The mammary glands make milk for babies, but from dogs to walruses, to chimpanzees, they only become obviously swollen during lactation. Outside that period, they shrink to a size barely different from males of the same species. This is expected if their only function is nourishment. Human females are different: the breasts become permanently enlarged at puberty. Some of this relates to growth of the milk-producing glands and ducts that carry milk to the nipple. But most of it is caused by the local deposition of fatty tissue.[15] Not only does this fat not produce milk, it creates a nuisance during movement (as witnessed by the strong sales of jogging bras). The only reasonable conclusion is that the fatty tissue that makes breasts permanently enlarged is a sexually selected ornament.

Our tour of sexual selection above implies that if women have sexually selected traits, then they are routinely subjected to male choice and/or used to engage in female-female competition. Consistent with the choice part, men are consistently attracted to breasts across cultures (who knew?). This preference is part of a larger search image that prioritizes what, for lack of a better term, could be called "curviness." Beyond the breasts, puberty also triggers gain of fat in the hips and buttocks in females, exaggerating a modest skeletal difference that widens the birth canal. When presented with silhouettes of female bodies, men from around the world consistently prefer a particular waist-to-hip ratio, of about 0.6 to 0.7.[16] The classic American pinup girl's measurements are 36-24-36, which requires prominent breasts and a waist-hip ratio of 0.67. While these proportions don't actually apply to very many women, it is apparently not an arbitrary cultural quirk. It is actually tapping into a long-standing male preference (one that also probably explains why anyone would ever bother eating at Hooters[17]).

What, if anything, might men gain from choosing curvy females? Most obvious is information about reproductive age. Curvy women are all adults, or nearly so, and thus potential mates. In addition, women (and men) generally fill in around the waist with age, causing their waist-hip ratio to increase to 1 (or more). So, the general male preference is specifically tuned to the bodies of young adult women. This is the counterpart to the information communicated to women by male traits that indicate attainment of maturity, like beards and broad shoulders.

Beyond general indicators of reproductive age, curviness indicates other things. Because milk production does not depend on the fatty tissue of the breast, it is not likely to be nursing ability in the narrow sense. However, the presence of fat in the breasts, hips, and buttocks does depend on being well-fed. A curvy female is thus likely to be not only of a suitable age but also in good condition. Living with (and doing my best to feed) my wife through her two pregnancies and many months of nursing, I can attest that the metabolic demands of maternity are impressive. It seems that, for millennia, men have been most attracted to women who are the right age to bear children and healthy enough to survive it. This is actually quite sensible. Over time, curviness became an asset in mate attraction, and over many generations it became a standard feature of the human female body.

The above indicates that humans have the maximally complicated form of sexual selection: males and females appraise each other and make choices (at least where such choices are allowed), and at the same time competition occurs within each sex. In other animals, mutual sexual selection is strongest where there are stable pair bonds. This is especially important for selection on the female traits. If males provide no long-term care for their mates or offspring, they have no incentive to be choosy in the first place. In chapter 12, we will see that men are, for the most part, more powerful physically than women. This suggests that in a world without guns or police, part of male choice and sexual selection of female traits was an exchange of beauty for the promise of long-term protection, for both the female and her offspring. While acknowledging that there are plenty of exceptions, this pair bonding feels like a norm (or at least an ideal) in our species. But is it? If it is, when did this arise?

The luminary of anthropology, Claude Lévi-Strauss, first argued that a common feature of all human societies is long-term pair bonding (marriage or its equivalent) combined with a set of social norms that promote reciprocal *exogamy*, or mate exchanges between focal families.[18] Lévi-Strauss believed this suite of habits to be a peculiarly human phenomenon but held that it likely developed many times independently as humanity spread around the planet from its African origin. In the half century since Lévi-Strauss put his idea forward, anthropologists made huge progress in determining the relationships of primates and in characterizing their social behaviors. In light of this progress, the University of Montreal's Bernard Chapais has made a compelling case not only for Lévi-Strauss being right about the combination of pair bonding and reciprocal exogamy as general and uniquely human but also that we should posit the simpler idea that it's universal because it is primordial.[19] That is, this goes back to the last common ancestor of present-day humanity.

Pair bonds can be monogamous and are in some other primates, but they don't have to be. Gorillas have a polygynous system with one dominant male that monopolizes essentially all females of reproductive age in the group. In some human societies, powerful men also work to create a similar arrangement. Chapais points out that what really makes humans different is that only humans use a *mix* of monogamy and polygyny. That is, the ability of a few men to have multiple mates does not eliminate monogamy in the same social group. This mix hints to a recent past in which polygyny may have been very common (thus explaining our residual dimorphism in male size), transitioning in part or completely to monogamy only in recent times.

Why do pair bonds evolve at all? Chapais suggests that because monogamy exists without any paternal care in some primates, it likely starts for reasons other than the joint rearing of offspring (although for our species this becomes important later). Rather, he argues that stable pair bonds likely begin as a form of mate guarding by males. Polygyny can emerge in small groups with big differences in power between senior and junior adult males. One silverback gorilla can monopolize all the group's females, and so he does. Meanwhile, all the other males lack mates entirely, with some dim hope of becoming the dominant silverback late in life. In contrast, monogamy evolves when males are not different enough in size to win a fight against another easily, so that a polygynous strategy is usually untenable. The strategy to gain a mate is thus thought to shift to picking a single mate and guarding her by being with her all the time.

Once a pair bond exists, however, it sets the stage for subsequent changes in parental care of offspring. As we became smarter, we also took longer to reach adulthood than our ancestors. Children were no longer independent at weaning—they were just helpless toddlers who eat solid food. Nursing suppresses ovulation in most mammals, including humans. But if weaning is nowhere near the end of child-rearing, a mother will find herself both nursing and looking out for toddlers and other youngsters as well. This is a full-time job, and help from a mate would greatly increase the odds of a kid surviving childhood. If most men are attached to one female, then their efforts benefit only their offspring, and the interests of the mates align. The persistent sexual ornaments of both men and women help reinforce the arrangement. Consistent with this, men set a higher appearance bar for prospective wives than for one-night stands.[20]

Although a scenario in which human monogamy emerges from polygyny in our recent ancestors is widely held by anthropologists, there are other, less common ways of arranging relationships that meet the same basic needs. In traditional societies in regions as diverse as the Tibetan plateau, the South Pacific, and Amazonia, a focal female may have multiple, publicly recognized male partners (that is, *polyandry*). This seems like the opposite of mate guarding, so the context in which it occurs has been studied for decades.[21]

In Tibet and South Asia, an acute shortage of agricultural land seems to be key. If offspring each form their own distinct family, subdivision would soon make each plot unsustainably tiny. To avoid this, a female's multiple husbands are often brothers, who thereby share a single parcel and whose total paternity is limited. This is a rather egalitarian way to achieve the same effect of giving the entire estate to the first- or last-born offspring, which was the norm in Europe.

From the standpoint of evolutionary theory, males ought to be more inclined to share a mate with a brother than a stranger because even the children they didn't sire are still nieces or nephews to them genetically. There are some societies, however, where polyandry is not fraternal. Several that are indigenous to the Amazon basin have an informal sort of polyandry that is not tied to land distribution. When a child is born, each of the mother's sexual partners is regarded as a co-father, and each has some material obligation to the welfare of the mother and baby. This is referred to as *partible paternity*.[22] While there can only be one father in the genetic sense, this arrangement is pragmatic. There is a tendency for such relationships to occur when individual males are absent for long periods and/or have unusually high mortality. Having multiple fathers thus becomes a sort of insurance policy.

CULTURE AND SEXUAL SELECTION

Male choice and female-female-competition are all around us. Still, it took evolutionary biologists and anthropologists quite a while to acknowledge it, both for animals in general and for humans in particular. The evidence could be dismissed as exceptional for animals and "purely cultural" for humans. However, the above discussion makes clear that our bodies are actually very much involved here, so we need to consider the possibility that behavioral phenomena related to mate choice coevolved with our dimorphic bodies and have been with us since we became human. In the last decade, interest in this topic has exploded,[23] and the case can now be made that modern interactions between and within human sexes (1) involve mutual mate choice and intrasexual competition and (2) probably aren't so different than those of our premodern ancestors.

Marriage or its equivalent creates a sexual selection dynamic that distinguishes us from even our closest primate relatives. Each individual is trying to secure the best mate they can, knowing that it will affect their long-term prospects. The choice of mate has implications for the future resources available to the couple and for the safety and well-being of the female, who is intrinsically vulnerable (see chapter 12). Unlike a winner-take-all situation, each person has a good shot at finding a mate. But there is a sort of fuzzy pecking order for each sex, and individuals have to figure out a strategy to "marry well."

Prospective partners presumably ask themselves a range of questions before they say, "I do," such as: Does he make me laugh? Will he be a helpful father? Does my dad respect him? Is she charming enough to forgo future dating? Does my mom get along with her? Does his or her family have resources? Throughout human history, who to take as a mate has often been the most consequential decision in a person's life. If this sounds like a Jane Austen novel, it's not a coincidence. Austen's stories brilliantly capture the dynamics of mate choice and the ways in which people navigate what they clearly recognize to be a high-stakes competition.

How do we humans compete for mates? The Canadian psychologists Maryanne Fisher and Anthony Cox have studied this in North American young people and have discerned four methods.[24]

1. *Self-promotion*. This covers several types of activities that are not only for securing a mate. For example, anything we do to enhance our perceived power and importance or our physical appearance is in this category. It also includes behaviors that increase our perceived autonomy, commonly termed "playing hard to get." Neediness is not attractive if one is trying to identify a partner, and aloofness (in moderation) can actually increase attraction. Directly attractive behaviors, such as flirting, again only involve the reaction of one person to another.

2. *Competitor derogation*. Mate competition is not just about pairwise interactions. This category includes anything one might do to put down others, either to a potential mate or to peers. According to Fisher's research, both sexes do this but in different ways. Men are more likely to insult a rival's physical strength or material possessions (e.g., their car). Women are more likely to put down other women's appearance. Female-female competition, usually in the context of a focal male, is also a common theme in popular song, presumably because it rings

true to fans.[25] This points to the notion that women believe men value appearance (and they generally do) and conversely that men believe women value the ability to protect and provide (and they generally do).

3. *Competitor manipulation.* If self-promotion as defined above is ethical and even somewhat healthy, this is a darker dynamic. We humans engage in deceitful conversations to reduce a competitor's interest in a potential mate, and if that doesn't throw the competitor off the hunt, we also resort to threats and coercion to desist.

4. *Mate manipulation.* This is essentially the human form of mate guarding, in which a person prevents access to their partner by potential competitors by sequestration or by never leaving their side. While many of us would see this as creepy and controlling, it happens to some extent everywhere, and is (or was) the norm in some societies. Present-day Taliban-ruled Afghanistan takes this to an extreme by requiring that all women must be accompanied by a male relative (*mahram*) whenever outside the home.[26] Islamic scholars have noted this practice was originally introduced centuries ago out of concerns for the safety of women and not to oppress them,[27] but the Taliban's other restrictions on females suggest it is part of a general strategy of control.

The above menu of manipulation represents a sort of tool kit of sexual competition, but any one of these strategies might fail for a given person. Some people simply couldn't care less about appearance, perhaps valuing compassion, creativity, or a sense of humor more. Others are put off by flirting or see right through attempts to be manipulated. Insulting a rival can reflect badly on the insulter by casting doubt about their self-confidence, which undermines the autonomy piece. The game, therefore, is complicated.

Fashion can be considered another avenue of intrasex competition and intersex signaling. While clothing is essential to survive the elements, our recently evolved visual orientation (see chapter 10) allows clothes also to form a canvas of self-promotion, whether to signal community identity, power, or (let's face it) *hotness*. For women, form-fitting gowns or jeans, plunging necklines, and midriff-baring crop tops help emphasize the underlying bodily cues that signal the alluring traits of young adulthood. Fit men can wear tight T-shirts or jackets that accentuate their shoulders, and optimal facial stubble management is a common topic in men's magazines. For those of us who are not model specimens, other adornments, such as jewelry, hair styling, and makeup, may do. As with purely biological traits, we can infer they are part of an extended form of sexual selection because they serve no obvious practical purpose, yet they occupy much brain space in the hours before a date.

Driven by our slow, resource-intensive reproductive strategy, we evolved bodily signals in response to sexual selection over hundreds of thousands of years. At a pretty early age, we develop awareness of their existence and of the power they can confer in certain situations. This can't help but bleed into others areas of life. For example, what attire is appropriate for work? Does it matter for my job if I get a big beer belly? One idealistic school of thought is that our bodies shouldn't matter in the workplace and that modest dress that deemphasizes secondary sexual traits is best if we want our colleagues to take us seriously. But the attractiveness of these traits is always lurking, and there is much evidence that it can be used to one's advantage. Most people use their available resources to compete, but some also opt out of the game. Either way, we are affected by it.

A recent study found that even among academic economists, a presumably fairly cerebral bunch, those rated as more attractive get better jobs and are more heavily cited than those who are rated as plain.[28] Catherine Hakim of the London School of Economics has found the benefit of being sexy applies to other fields as well, and in a 2011 book, she dubbed it "erotic capital."[29] Hakim argues that we may do well to acknowledge its existence and stop trying to suppress it, with the expectation that it will help women in particular achieve professional parity with men. The argument is logical, but one can also see the danger of encouraging objectification in a deeply patriarchal world. Nevertheless, in every school, factory, farm, and office, we are making often unconscious calculations about whether and how to exploit our erotic capital.

CHAPTER 12

Sex Versus Sexism

Synopsis

Measurements of human sexual anatomy (with rare exceptions) fall into male and female bins, but other widely accepted sex differences do not. While sexes are real and useful categories, we must be careful not to equate overlapping distributions that have different averages with categorical binaries.

Content warning: This chapter discusses statistical aspects of rape and domestic violence.

—————◆—————

It has become somewhat fashionable today to question whether our species really does exhibit an essentially binary sex arrangement.[1] There is certainly something noble behind the urge: there has been far too much suffering for far too long by people who don't fit comfortably into some or all of the norms expected of people born into a male or a female body. So, by all means, let's storm the castle and stop the hating. But does that mean there really are no such things as sexes? Have we been under some collective delusion for millennia, as some would suggest?

In this chapter, a distributional view of sex differences is presented that underscores why the notion of a binary persists: there are a few traits that really do rather cleanly separate humans into male and female piles. (Your grandmother is not completely nuts.) However, this binary situation applies to far fewer traits than is commonly understood. Traits described as "sex differences" often have largely overlapping distributions in men and women, with consistent but only modestly different *averages* in male and females. The confusion of effectively binary traits with those traits for which males and females largely overlap, but differ *on average*, leads to a whole host of generalizations and misstatements. This is made worse when traits put forward as characteristic of one sex or the other have not been (or cannot easily be) measured. Further, even sex tendencies that can be measured can sometimes be explained by societal biases and upbringing (and so may disappear if those biases are rectified). In short, while *sexes* are real, little else is similarly binary, and thus this reality cannot justify *sexism* in society.

SEX DIFFERENCES VERSUS SEX TENDENCIES

"Congratulations, it's a _____!" Some version of this declaration accompanies the birth of nearly all humans, whether it is issued by an ultrasound technician, a midwife, an obstetrician, or a panicked partner in the back of a Buick station wagon. A quick inspection of the private parts of the fetus (after about twenty weeks) or newborn makes such a declaration a simple task in nearly all cases. As was discussed in chapter 2, the distinct male and female genitalia develop from common embryonic precursors, mostly in the first trimester of gestation. Do we end up with a continuum or something more distinct? We can measure the male and female parts that correspond to the primordial genital tubercle, the penis and the clitoris, to get a sense of how different they are in the end.[2] Consistent with your intuition, it is apparent (see figure 12.1) that the distributions do not overlap. Similar nonoverlapping plots would result from measuring the vaginal opening (the equivalent of which "zips up" into a tiny slit during the formation of the penis) and traits related to the ovaries and testes. The most extreme cases are traits that are truly sex-specific, such as production of sperm versus eggs (which

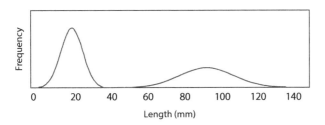

12.1 Distribution of the length of the derivative of the genital tubercle in adult humans.[2] Clitorises fall on the left, penises on the right.

defines sexes for zoologists—see chapter 3), the uterus (female—see chapter 2), and the prostate gland (male). If we were to present these as a distribution, they would have all members of one sex piled up at the zero point and the other sex forming a distinct cluster of nonzero values. I will refer to both kinds of nonoverlapping distributions between sexes as *sex differences*.

I chose to begin with the reproductive traits that define sex in the minds of most people, traits that really are binary sex differences in their essence. But how many other traits are there like this? It turns out there are few. For example, it is often said that "men are taller than women," and *on average* that is true. And when someone says, "Men weigh more than women," there is a real sex effect there. But if we reflect on our own experience, we also know tall women who tower over short men, and big women that weigh much more that most men. Putting height and weight onto the same common plot as we did in figure 12.1 for genitalia, we get a very different impression (see figure 12.2). A woman of only average height is taller than several percent of men, and a man of average height will be shorter than several percent of women. So it is not true to say that "men are taller than women" as a generality. Such average effects should be treated very differently than sex differences, so here I will refer to them as *sex tendencies*.

The fact that height and weight are not binary sex differences is important to understand, but such nonbinary distributions can lead to striking categorical differences at the extremes. Take, for example, basketball teams. There is an advantage to being tall in the game, and both men's and women's teams reflect this. But if we look at how the height distributions relate, we see that at the very tallest end of the complete distribution, there are only men. This is one reason why having separate men's and women's basketball teams makes sense: in the most elite competition: women will be at a consistent disadvantage otherwise (see the epilogue).

In a sport in which being tiny is optimal (e.g., horse racing), the very tiniest people will all be women. The rarity of female jockeys is thus surprising: they should actually dominate. Yet the first woman to race in the Kentucky Derby, Diane Crump, didn't race until 1970, and the first in Britain's Grand National did not race until 1977. This was because women were systematically denied jockey licenses in the U.S. industry. That is, until Kathy Kusner successfully sued the Maryland Racing Commission in 1968. In the United Kingdom, a policy forbid women jockeys until 1975.[3] Although a blogger affiliated with Churchill Downs

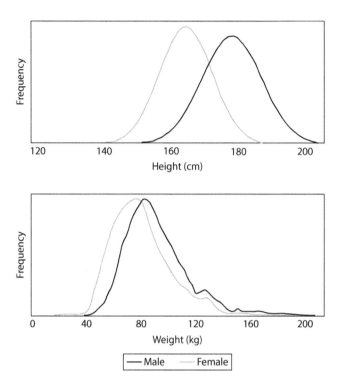

12.2 Distributions of height (top) and weight (bottom) in adult men (black line) and women (gray line). Height data are for North America, Europe, East Asia, and Australia, while weight data are for American adults.

Source: Height data compiled by Roser et al., "Our World in Data" (2019), https://ourworldindata.org /human-height. Weights plotted directly from the sample data summarized in the National Center for Health Statistics, *Anthropometric Reference Data for Children and Adults: United States, 2015–2018* (2021).

opined that "there's no noted reason for why the sport is dominated by male jockeys," the only viable candidate reason is historical sexism, pure and simple.[4]

INTERSEX AND THE BINARY

The above definition of sex differences does not deny the existence of intersex people with atypical sexual anatomies. Their differences of sexual development (often referred to as disorders of sexual development [DSDs] in the medical literature) are very real, and our social, medical, and legal systems need to acknowledge them in a compassionate way. Does the existence of intersex people mean that the apparent binary aspects of human sexes noted above are actually useless rubbish? I argue it does not.

To be a "sex" in the biological sense, a form needs to be anatomically coherent and reproductively functional, and comprise a substantial fraction of a species' individuals. In the context of figure 12.1, there are two, with the population forming the hump on the left being female, and the one associated with the right hump being male. If sex were a continuum, the humps would be connected by a large fraction of individuals with intermediate values, or there would be no humps at all. The fact that this is not the case makes sense in light of what we learned in chapter 7: internal fertilization requires complementary anatomies. This selects for genetic systems (discussed in chapter 5) that produce distinct alternative developmental outputs. If there were an additional sex, it might produce a visible third hump in the distribution. As we will see in chapter 13, some animals do have such third forms, most often two types of males with very different mating strategies.

Deviations from typical sexual development that produce an intersex result are actually a collection of diverse traits. They often cause sterility in the absence of medical intervention and are rare. For example, genitalia that are not clearly male or female occur in roughly 1 in 5,000 births.[5] This is similar in frequency to spina bifida, the neural tube closure birth defect, which affects 1 in 2,500 births. In most (but not all) cases, intersex people harbor a chromosomal aberration or a mutation that disables a gene regulating development of the gonads or genitalia, or a gene essential for hormone signaling.[6] For all these reasons, such conditions are rightfully considered to be medical concerns, although beyond infertility they often do not present general health burdens.

The previous paragraphs make the case for the validity of male and female biological sexes in our species, but it could be seen by some as justification for something darker. I therefore need to be equally clear about something else: for intersex people, not being forced to conform to one or the other standard-issue sex (either anatomically or legally) is a matter of basic human rights. Society really has no justifiable interest in regulating how such people choose to manifest with regard to sex and gender. Life is hard enough without adding that kind of invasive scrutiny, and it should be left to the patients, families, and doctors involved. My point is simply that it would also be a distortion to consider intersex people as evidence for a meaningful continuum of sexual development. To acknowledge them, we might more precisely describe human sexes as *effectively binary*, or *binary**. This is preferable to blowing up the concept of biological sexes, which is neither practical nor likely ever to be accepted by society.

DUBIOUS TENDENCIES

It would greatly simplify life for all humans if we could say (and be correct) that "men's brains are like this, and women's brains are like that, and as a result there are many sex differences in mental abilities that we can use to structure society." Such arguments go back hundreds of years, and even today a sizable collection of pop-science books attempt to make just such an argument. From empathy to spatial reasoning, from aggression to our inclination to read a story with our child, our internal drivers and abilities have been put forward as bona fide sex differences. When subjected to scrutiny, however, there is no evidence that any of these represent sex differences and only weak evidence that they even represent innate sex tendencies. As this has been the subject of several recent critiques by highly qualified authors, I will only summarize key points here.

In her book, *Delusions of Gender*, Cordelia Fine meticulously rips purported mental sex differences to little bits, and gives us good reason to be skeptical of claims for their existence.[7] Bona fide sex differences, such as fetal testosterone, have been appealed to as a plausible mechanism for creating a distinct "boy brain" and "girl brain," yet there is no clear indication that the anatomy of the brain is in fact substantially affected by it. Muddying the waters further, Fine marshals considerable evidence that, as social beings, we care deeply about our relationship to others and begin tweaking our self-conception and choices in response to gendered feedback essentially from birth. This tends to enmesh the origins of any measurable sex tendencies with preexisting social forces, and often these forces are sufficient to explain the tendency. Not surprisingly, then, human behavioral outputs are not cleanly aligned with our hormones.

I will not go so far as to argue that there are *no* behavioral sex tendencies. For starters, biological sex is highly (but imperfectly) predictive of sexual orientation: Women are much more likely to be attracted to men than to other women, and vice versa (though see chapter 13 for consideration of those who don't fit into this). It is highly unlikely that this is a learned trait. In addition, biological sex seems to be a direct influence on childhood play, where even very young toddlers (presumably too young to have learned the behavior) show a reproducible sex tendency (but not binary sex difference) toward certain toys and modes of interacting with their peers.[8] That is, little boys really are more likely to tackle each other and play with trucks than are girls, who really are more likely to play with dolls. Studies of girls with unusually high testosterone levels suggest that hormones may influence these tendencies.[9] Given that play is at least in part a form of rehearsing adult roles, evidence for a similar sex tendency in other primates, and the history of sexual selection in our species' recent history (see chapter 11), this is a plausible albeit still rather mysterious phenomenon. Even here, however, cross-typed behavior is common, and children explore the cross-typed toys more when they are away from the public setting of the laboratory. It appears that even very young children have internalized a sense of what they are "supposed to do" when someone is watching, and this exaggerates any innate sex tendencies.

Neuroscientist Daphna Joel and science writer Luba Vikhanski offer a framework for comparing the brains of men and women in their 2019 book, *Gender Mosaic*.[10] An expert in brain imaging, Joel has examined patterns of activity in large numbers of human volunteers. Many studies have reported sex tendencies in such data, tendencies that can be identified if we take lots of measures of brain activity from many people. We have no reason to believe these data are intrinsically unreliable. But does this mean that, within individual brains, all tendencies align in the expected way with the subject's gender? Joel's research has shown that they most definitely do not: for any one person, there is a unique mix of "masculine" and "feminine" tendencies. Only when many traits are examined as a group does any consistent sex effect appear, a disproportionate sorting that is disconnected from any one mental circuit. This means that generalizations about any one of the measured traits will always be mistaken. And if we connect this work to the arguments of Fine, even this meta-pattern is likely influenced by social feedback given the subjects were adults living in a highly gendered world.

Premature claims for strict sex difference can even touch nominally objective biological research. As we saw in chapter 4, genetics has made great strides in clarifying how sex chromosome differences allow distinct male and female forms to emerge from a single species. The alternative sex chromosomes of males and females would seem to be the safest sex differences of all. But even here there has been a tendency to ascribe all the power of the sex-specific genotype to a single, most-upstream "master regulator." As Harvard's Sarah Richardson has noted, the notion of *SRY* as the sole determinant of sexual differentiation collapsed soon after its initial discovery.[11] The *number* of X's is also very important to the development of male and female. This was initially unexpected for a good reason: one of the female mammal's X chromosomes is largely silenced, forming a highly condensed structure called the Barr body.[12] An inactive X would seem unable to have an impact on anything and led to the expectation that the Y is the sole driver of sex. However, it eventually became clear that not all sites are silenced in the Barr body and that the dosage of a few X-linked genes matters tremendously for sex.[13] From a comparative perspective, the dual control by the Y and X dosage is not surprising given that X dosage alone dictates sex in other animals, such as fruit flies and roundworms.

Beyond the models of biologists, sex chromosomes have also been proposed to be largely determinative for more than just sexual anatomy: binaries at the genomic level are said necessarily to lead to similar binaries in behavior, cognitive abilities, and emotions. Geneticists, journalists, policymakers, and activists are all people, people who were (and remain) subject to biases. When science enters the public domain, simpler messages are easier to sell than complex ones, although the latter are usually more accurate. Even among those working in the field of contemporary medical genetics, many of whom are women and self-described feminists, the tendency to regard sex as consistently creating categorical *differences* is a tempting shorthand when, in reality, most attributes reflect sex *tendencies* instead. Richardson warns that "using sex as a principal category for biological analysis may cause important differences within groups to be overlooked. This may lead to the use of therapies that do not provide generalized benefits to sub-groups." She has a good point. For example, women are often said to experience a different spectrum of heart attack symptoms than men, but the actual numbers indicate that both sexes experience all symptoms, and the sex effect is a modest tendency.[14]

THE BIOLOGY OF VULNERABILITY

The previous section can be boiled down to this: beware of simplistic statements ascribing fixed differences to men and women. You may be feeling a bit confused, however, in squaring that with the world in which we live. Males are overrepresented in the most lucrative and powerful positions in society, and it's not hard to find people happy to argue that this is actually the way things should be. The history of my own country is littered with explicit exclusion of women from influential institutions and the more lucrative professions. My current employer, the University of Maryland, College Park, didn't admit female undergraduates until 1916, and it would take Yale and Princeton another fifty-three years to do the same. When I was growing up in the 1970s and 1980s, I did not initially question why my father's fellow well-paid engineers were uniformly men, while the clerical staff that performed different, but equally essential tasks (at much lower salaries) were nearly all female. We clearly live in a gendered world.

The reality of patriarchy begs a fundamental question: if there are so few clean differences between male and female, then from where exactly does male power come? This could quickly get us into a very complex discussion of culture and history, one better led by others. But here I suggest we cannot ignore that a historical (and ongoing) root of historical male power comes from its most literal, biological form: muscles.

We have some sense that men, at least on average, are stronger than women. Strength is tightly dependent on skeletal muscle, whose contractions move our arms, legs, and every other part of our body we can control voluntarily. But is this a sex *difference*, or a sex *tendency*? When I began looking into it, I assumed it would be the latter, as for height and weight. To my surprise, however, it is actually much closer to a sex difference, like genitalia. As shown in figure 12.3, when

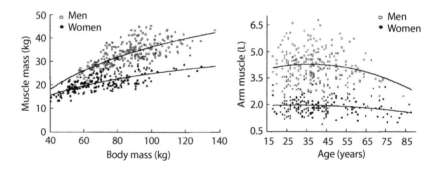

12.3 Female versus male skeletal muscle mass. Left: Total muscle masses plotted against corresponding total body masses. Right: Arm muscle volume plotted again the age of the person. In both plots, the lines show the regressions (i.e., best-fit relationship) of the muscle distribution as a function of body size (left) or age (right) for each sex.

Source: Data modified from I. Janssen, S. B. Heymsfield, Z. M. Wang, and R. Ross, "Skeletal Muscle Mass and Distribution in 468 Men and Women Aged 18–88 Yr," *Journal of Applied Physiology (1985)* 89, no. 1 (July 2000): 81–88, https://doi.org/10.1152/jappl.2000.89.1.81.

Ian Janssen of Queen's University, Ontario, and his colleagues measured skeletal muscle mass in hundreds of women and men, they found remarkably little overlap. This is true even when differences in total body size are taken into account. See, for example, how in the left panel of the figure that, for any given body mass, nearly all the male points lie above the largest female values (especially for those who weigh more than 80 kilos, or about 175 lbs). This difference is body-wide, but more extreme in the upper body, such as the arms (right panel).

The implication of Janssen's muscle measurements is sobering: even a relatively wimpy guy (like me) has more upper body strength than almost any woman, even if she is taller, heavier, or substantially younger. This is a basic feature of humanity, and its developmental origin is clear: skeletal muscle growth is strongly enhanced by testosterone. Blood testosterone concentrations in men are roughly ten times higher than in women, with no overlap in the two distributions.[15] The fact that testosterone and the upper body strength it confers are true sex differences has some uncomfortable implications. In particular, without the aid of a weapon, women as a group are vulnerable to being overpowered by men. Thankfully only a small fraction of men actually act on this, but at some level, the mutual understanding that men can physically dominate women conditions our everyday lives. The existence of women's shelters and, on my campus, the telltale blue-lit emergency stations and night shuttle buses, are all acknowledgments of this vulnerability. It is always there, even in calm situations with no hint of conflict.

Sometimes there is conflict. Women make up 94 percent of reported rape victims,[16] and 92 percent of those convicted of sexual abuse are men.[17] Women do sometimes force men into unwanted sex, but the asymmetry in numbers and degree of violence is striking.[18] And then there is homicide. Of the roughly five thousand people convicted for homicide in the United Kingdom from 2010 to 2020, 92 percent were men (the fraction is similar in the United States).[19] Two-thirds of homicide victims are also men, so men are at greater risk of both committing and falling victim to homicide. If we use these numbers to predict the fraction of homicides that are men killed by women, and vice versa, we expect that more women are killed by men ($0.92 \times 0.33 = 30\%$) than vice versa ($.08 \times .67 = 5\%$). However, there are indications that there are other forces at play beyond a general tendency for men to be more violent:

- According to data from the Federal Bureau of Investigation (FBI) data,[20] American men killed roughly 4.4 women for every man killed by a woman, pretty close to what is expected from the 1:6 expectation above. But when the U.S. cases are trimmed to just those who killed without a weapon (i.e., using their own hands), the ratio inflates to nearly 1:15.

- Eighty percent of women in the United Kingdom who died in domestic violence incidents were killed by a spouse or partner (who we can infer are mostly men), but only 39 percent of male domestic violence victims were killed by a partner or spouse (men were more likely to be killed by other family members).
- Women murder victims were three times more likely to die by strangulation or asphyxiation than were male murder victims (19 percent versus 6.5 percent). Because most murderers are men, this means men are more likely to strangle a female victim than a male victim. Recent numbers from Fairfax County, Virginia, confirm this: of 378 cases of strangulation (most nonlethal) reported from 2021 to 2023, 362 of the assailants were men.[21]

The morbid details above make this point: the strength that men possess doesn't just get used for unspoken intimidation of women: it is used to hurt or kill them often enough that the potential is likely obvious to everyone. In her landmark study of rape, *Against Our Will*, Susan Brownmiller argues that genitalia are at the heart of female vulnerability:

> Man's structural capacity to rape and woman's corresponding structural vulnerability are as basic to the physiology of both our sexes as the primal act of sex itself. Had it not been for this accident of biology, an accommodation requiring the locking together of two separate parts, penis into vagina, there would be neither copulation nor rape as we know it. . . . From prehistoric times to the present, I believe, rape has played a critical function. It is nothing more or less than a conscious process of intimidation by which all men keep all women in a state of fear.[22]

While our genitalia do create a penetrator and one who is penetrated, we could rephrase this just as accurately as the "engulfed" and "engulfer," casting the penis in the more passive role. Figure 12.3 suggests the more fundamental difference at work here is not genitalia but strength. Without it, males could not force women to do anything. Most men are neither rapists nor murderers, but the power they have enjoyed throughout history stems in considerable part from awareness of this fact.

CHAPTER 13

Queerly Normal

Synopsis

If evolution selects for reliable reproducers, why are so many humans gay and queer? Several lines of evidence suggest that exclusive homosexuality, especially in men, represents an environmentally controlled alternative state, or polyphenism. By increasing the survival of related children, homosexuality likely more than makes up for the reproductive sacrifice it imposes.

———•◆•———

On June 26, 2015, my daughter and I were at the Ronald Reagan Washington National Airport, munching our lunches at the iconic Ben's Chili Bowl. The TV was tuned to CNN and was delivering breaking news. CNN is always delivering breaking news, of course, and at first I did my best to tune it out. In this case, however, it was hard to do: the U.S. Supreme Court had just ruled, in *Obergefell v. Hodges*, that all states must recognize the rights of same-sex couples to marry. As I tried to process the news for my nine-year-old, a family with children came in and sat down next to us, largely unremarkable but for their colorful pride shirts. Confident they would be pleased, I turned to them and pointed out what I was hearing on the TV. "Do I see the news?" the mom asked in happy exasperation, "We were plaintiffs in the case!"

The smiling faces of April DeBoer and her family said much about the significance of that day. Indeed, the recent gains in civil rights by gay Americans have been astonishing to those who both advocated for them and opposed them. As more and more people and states stepped forward, we seem to have finally admitted to ourselves that every family, neighborhood, and workplace can count its queer members and that this was not going to change. The click of a cultural ratchet was heard as a wave of people came out and increasingly took their rightful places in public life. The 2022 passage of the Respect for Marriage Act appears to have finally settled the marriage issue, although not everyone (straight or queer) aspires to being married.

Before we run a smug victory lap, however, we can (and should) hope for much more than legal acceptance. Race provides an informative parallel. While some do explicitly advocate for complete separation from racial and religious minorities (I'm looking your way, Scott Adams), the vast majority of white Americans have big enough hearts and rational enough minds to accept our nation's diversity. As with gay marriage, civil rights laws have been key players in supporting the emergence of a more tolerant society. But alongside acceptance is the persistent feeling among many of European ancestry that they are still the "all-American" norm. The implication, often tacit, is that anything else is "other": abnormal and somehow less American. Othering is a pillar of racism and can lead to suspicion, harassment, and sometimes murder.

Despite the harm it causes, this persistent othering is not actually synonymous with malice. It manifests across the entire political spectrum, from white supremacists obsessed with racial purity to *Prairie Home Companion* fans who feel a deep commitment to social justice. For a more durable form of justice to take root, the kind that can dismantle segregation, income inequality, and police brutality, othering must somehow be transcended. Science is playing a central role here.

Human genetics research has efficiently demolished the reality of races as distinct types. Mirroring earlier paleontology discoveries made in the Rift Valley, it has shown how our species arose in East Africa. To this day, this region remains the most genetically diverse, despite the tendency of non-Africans to label all its many peoples as simply Black. European, Asian, Australian, and Indigenous American populations represent partially overlapping subsets of that African diversity, as expected when small groups of founders leave a source population. From this we see the truth: regardless of where we live or what our skin tone is, we are all basically Africans. Some out-of-Africa migrants did acquire a tiny number of adaptive mutations that promoted vitamin D production in cold climates.[1] One hopefully sees rather quickly that the resolution of a historical vitamin deficiency provides a poor basis for a claim of innate superiority. By revealing the shaky premises upon which the notion of race is based, science undermines othering and supports the slow march toward real justice.

Vis-à-vis queer people, I suspect most straight people inhabit the same place of "acceptance with persistent othering" in which much of white America resides with respect to race. Despite increasing legal and social acceptance, the sense of same-sex orientation as an oddity, as grounds for being deemed other, remains largely unopposed by science. Absent credible explanations for why it exists, even in tolerant societies gay-straight relationships are stunted and homophobia persists. "Gayborhoods" may be fading in the online era, but the American tragedies at Orlando's Pulse nightclub and Colorado Springs' Club Q reminds us there is still plenty of homophobic hatred in the United States. In other countries, the situation is much more dire. Worldwide, both state-sanctioned and less official repression of gays continue to kill, imprison, and marginalize millions of people.[2] That the victims pose no actual threat to the regimes that torment them indicates that fear of the other makes them easy scapegoats.

Is there a role for science here? Just as genetics and physical anthropology are helping kill racism, might consideration of the biology underlying queerness undermine homophobia as well? I believe the answer is almost certainly yes, yet surprisingly there is not (yet) a widely accepted model for why and how homosexuality exists. Were it to be found, however, it could have the power to de-other gays, reduce homophobia, and even save lives. A fuller understanding is therefore very much worth seeking.

THE EVOLUTION QUESTION

It is now clear that the roots of a homosexual adulthood often emerge in very early childhood and are highly resistant to disincentives. To paraphrase Lady Gaga, gay people really are born that way (or at least become that way well before puberty). The immediate importance of this fact is in revealing the futility of "gay conversion therapy" and other attempts to coerce people into straightness. However, if gay orientation is innate, then it is a feature of the human organism that develops, just like handedness, height, and hair color. It must also have some sort of evolutionary history. This puts us squarely in the realm of biology.

The basic logic of biological evolution is pretty simple: The future belongs to those that manage to survive and propagate the next generation of their kind. For some organisms, such as creosote bushes and sea squirts, that propagation can include asexual processes like budding. For us humans, however, we're talking about babies. I hope that I'm not telling you anything new when I point out that (at least until very recently in a few privileged places) making babies absolutely requires men and women having sex with each other.

Although you probably already knew where babies come from, an intriguing mystery remains: Across world cultures, a substantial proportion of our species strongly prefers homosexual relations over the kind of straight sex that makes babies (see below for details). This begs the following question: If evolution selects for reliable reproducers, then why is homosexuality so common? The answer to this question is not immediately obvious, nor is it agreed upon by experts who have pondered it. Because homophobia remains and people still die of it, however the answer really matters.

To understand fully how a trait evolved, we need information on how it affects reproduction and survival of both the affected individuals and their relatives, and an understanding of how the trait is produced. Having spent most of my research career working with tiny roundworms with a two-day life cycle, I can say with some authority that we humans are lousy research subjects. Such long generation times! So many ethical constraints! At least there are a lot of us, and we have no shortage of curiosity about our own biology and habits. As a result, despite our shortcomings as subjects of experimentation, science has provided some important clues about the development and evolution of sexual orientation.

A BRIEF NATURAL HISTORY INTERLUDE

In the animal world, we generally think of males as one thing, and females as another. However, creatures as diverse as dung beetles and salmon produce alternative forms of a given sex, a sexual *polyphenism*. In one type, males either mature into large, territorial forms with sexually selected traits related to courtship and intermale contests (like the bright tail of a peacock, big horns, etc.), or they mature into a smaller form that lacks flashy male traits but is nevertheless fertile. By resembling juveniles or females, they fail to elicit a potentially injurious challenge from large males and thereby gain access to an alternative path to offspring.

Atlantic salmon provide an example of this kind of sexual variety. Most males spend years at sea, after which they return to the stream of their birth to mate. A subset of males, however, are small, reach sexual maturity early, and never leave their home stream. If the only route for landing a mate were the traditional option of combat, courtship, and territory defense, these runty preemie males would be evolutionary toast. However, there is a method to their apparent madness. Far from being disinterested in sex, they move undetected among females and carefully observe the courtship around them. When a female finally releases her eggs, they dart into the mating arena and quickly release sperm.

Were precocious males able to sire as many progeny as the resident territorial male, this strategy would seem to be very appealing—all the payoff with less risk of death at sea or in the jaws of a hungry bear during the spawning run. However, a precocious male sires only about a tenth as many progeny as a successful territorial male.[3] Why would fish opt for such a crummy outcome? Most likely because dead fish father no babies. With an estimated mortality of about 10 percent per *month* spent at sea, after two years, only 7 percent of fish that reached the ocean are still alive to make the return trip (which is itself fraught with dangers). This starts to make the fertility achieved by precocious males seem like not such a bad deal.

Precocious males trade high-stakes competition, which either works spectacularly or fails completely, for a lower-risk strategy that has a smaller maximal benefit but is more likely to produce a nonzero benefit. To put it mathematically, they are emphasizing their geometric mean rather than the arithmetic mean (or average) over the years. Add to that ongoing human fishing pressure on large adults, and we may expect to see more and more males adopting this strategy. Such a sexual polyphenism is an *alternative reproductive tactic* (ART), a stable path to fertility that has evolved in unrelated animals independently. For example, the University of Montana's Doug Emlen and Indiana University's Armin Moczek and their colleagues have studied dung beetles with intense male-male competition for mates. In some species, larger males have big horns that they use in dominance contests with other males, while a significant fraction have no horns. Under the cover of this camouflage, they engage in "sneaky" mating tactics reminiscent of precocious male salmon. Natural selection has thus led to a plan B for some males that still allows them to reproduce.

The salmon and beetle ARTs reduce reproductive success relative to territorial males, but they still depend on fatherhood. In highly social species, however, some individuals have no offspring at all. For example, a honeybee hive has only one reproductive female at a time, the queen. That doesn't mean she is the only genetically female bee in the hive—thousands of her sisters toil on her behalf. These sterile workers are not distinguished from the queen by their genes—they were simply reared without royal jelly. As adults, the development of their ovaries is further suppressed by pheromones secreted by glands in the jaws of the queen.[4] By pooling their efforts, bees can adopt a lifestyle that would be impossible for solitary bees. The abdication of reproduction by many individuals within a group of related animals is not restricted to invertebrates. The somewhat grotesque naked mole rats of Africa are highly social, and as with honeybees, the fertility of other females in the colony is suppressed by the queen (specifically by pheromones in her urine).

One cannot understand the origin and frequency of alternative reproductive tactics without awareness of the suite of factors that dictate success for each form and/or the groups to which they belong. For example, the sneakier male alternative needs to remain rare in order to be a viable strategy: if no male fish committed to the sea migration and development of sexually selected traits that entice females into laying their eggs, then there would be no eggs to fertilize. Similarly, sterile workers form the majority of the workforce of a bee hive or naked mole rat colony, yet they support the success of their close family member. In doing so, their genes are passed on indirectly, further reinforcing the strategy. It would be a mistake to call these alternative morphs "abnormal"—they exist for good reasons.

THE OPTIONS

The plausible explanations for why homosexuality exists in *Homo sapiens* can be seen as the four possible combinations of two different attributes, as shown in figure 13.1. Each attribute could be a continuum, but for simplicity we can initially consider their extremes. One of the attributes is the nature of homosexuality as a biological phenomenon. It could be a polyphenism (or set of polyphenisms), similar to those described above. If it is a polyphenism, then we should regard traits associated with it to be coordinated in a way that is somehow adaptive. Or maybe homosexuality is not a polyphenism but instead simply reflects an incoherent suite of variations that are tolerated in our species for some reason.

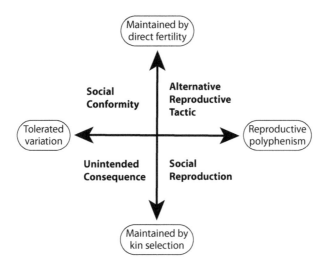

13.1 Combinatorial hypotheses for the evolution of human homosexuality.

The other attribute is the evolutionary force that is promoting homosexuality. One extreme would be direct reproductive success (i.e., parenthood). Although contrary to intuition, perhaps during most of our species' history people attracted primarily to their own sex did not actually have reduced fertility. If this were true, then the question posed at the outset essentially disappears because it is based on a fallacy: same-sex orientation and/or activity did not historically reduce individual reproduction. Or maybe being gay did reduce fertility, but it was maintained as an unintended by-product of a genetic makeup that often benefited straight relatives. This indirect effect via the success of close relatives is referred to as *kin selection*. In essence, kin selection works because the sacrifices it promotes in some individuals are more than compensated for by the enhanced success of close relatives that carry the same genetic variants. This is thought to underlie many cases of suppressed fertility among social animals, as in bees and mole rats.

The pairwise combinations of these two attributes cover the space of plausible explanations. When homosexuality is maintained by high rates of reproduction, this could be in the context of socially enforced conformity (for example, a closeted bisexual or gay person in a straight marriage), or as a polyphenism that constitutes an ART. If homosexuality is instead maintained via kin selection, again we have two ways in which this could manifest. It could represent a form of nonreproductive specialization, like that of other animals with strong social control of breeding. Alternatively, low reproduction of gay people may be a by-product of a system that confers straight relatives with excellent fertility. While social reproduction proposes the existence of one or more nonreproductive "castes" with a particular role to play, kin selection through enhanced fertility of relatives does not require such a role. Instead, the specific traits of gay people would be of no particular importance, as if they were innocent bystanders whose reduced reproduction exists solely because of the superb fertility of their straight relatives. I term this the "unintended consequence" scenario.

Each of these four broad options has something going for it, but each also has weaknesses. They are also not entirely mutually exclusive. Let's examine them and see where we get.

Option 1: Social Conformity

This scenario bears in mind the fact that most of our sexual activity does not result in baby making (thank goodness—otherwise, the word would be a very crowded place). Much of it can be considered "sexual play" that feels good and creates bonds between us. This nonreproductive aspect of sex is also observed in some of our closest relatives, such as pygmy chimps (bonobos) and gorillas,[5] and greatly predates the origin of *Homo sapiens*. Maybe the variation in what we like to do and with whom we like to do it has little or no historical impact on our likelihood of biological parenthood. If this seems implausible, imagine a woman in a premodern society that is pregnant for nine months and then nurses for two years (largely suppressing ovulation). Around the time her toddler is weaned, she has sex with a man a few times, she gets pregnant again, and the cycle repeats. In such a situation, she could have ongoing relations with another woman, and even feel most fulfilled with this partner, yet have fertility identical to one who did not. A comparable scenario for a gay man could also hold—occasional sex with a female partner could produce numerous offspring, even if he frequently partners with another man as well. We therefore need to consider the extent to which gay people have kids.[6]

For a variety of reasons, estimating how often gay people have biological offspring is difficult to do with great precision. For starters, it's not clear who should even count as "gay." Should sexual orientation be defined by what someone finds most arousing or by what they actually do day to day? How should we classify people who have sex with both men and women over the course of their lives? How about someone who has a strong same-sex attraction but is in a straight marriage? Alfred Kinsey first noted that there is a continuum between exclusively straight and exclusively gay, and he measured it with a seven-point scale, but even this score might change over one's life. Despite these challenges, in the early twenty-first century, many Americans are comfortable declaring themselves to be gay or queer, so this provides at least a provisional classifier. A report by the Centers for Disease Control and Prevention's (CDC) National Center for Health Statistics found that over 90 percent of self-identified bisexual and lesbian women, and 70 percent of gay men report having had at least one opposite-sex encounter in their lives.[7] These kinds of encounters presumably make babies.

Does "straight dabbling" actually lead to substantial gay parenthood? Apparently yes, at least in the United States. A U.S. Census Bureau report found that 19 percent of same-sex households surveyed had children present, 84 percent of which included biological offspring of a parent.[8] By comparison, 44 percent of U.S. households headed by straight couples had children present, 94 percent of which included biological offspring. By surveying participants in gay pride events, researchers in Chicago estimated that 17 percent of gay men were fathers, compared with 60 percent of straight men. Artificial insemination and other assisted reproductive technologies add some uncertainty to these numbers, but these practices remain rare enough that the basic conclusion still holds: gay people can and do have plenty of biological children, though not as often as straight people.

So is the reproductive success favored by social conformity enough to explain the persistence of homosexuality? Probably not. For starters, if being straight is the adaptive norm for our species, and people with same-sex attraction were occasional aberrations, then we would expect the distribution of sexual orientation to have the fewest people at the completely homosexual end of the spectrum. Does it?

In his groundbreaking survey work, Kinsey's scale (where completely heterosexual was rated zero) allowed people to identify their orientation anywhere along a gay-straight continuum as they saw fit.[9] The fact that this sort of survey result emerged in 1949 America, when open homosexuality was not acceptable, is remarkable. Kinsey did his best to generalize results from his survey of thousands of white American men to men overall by taking into account the age, marital status, and education of the U.S. population as a whole. Although Kinsey did not present his data in a way that highlights it, the most common nonzero score for men was completely homosexual, with relatively few identifying as bisexual (figure 13.2).

Kinsey's findings for men have been generally supported by subsequent work in the United States and in other countries. In more recent sexual orientation surveys, however, an "almost straight" orientation is more common than in the earlier work. In fact, as is evident in the CDC data (see class 1 in figure 13.2) and as was noted by the University of Utah's Lisa Diamond, of the men who have experienced or considered same-sex sexual behavior (SSSB), those who are largely heterosexual outnumber those who identify as exclusively gay.[10]

Kinsey produced a vast trove of data on women, but he didn't generalize his numbers to the overall female population. More recent studies do give us such an overall picture (figure 13.2), and they consistently support two conclusions. First, women are much more likely to identify as bisexual than men. Second, the strong peak in the distribution at the exclusively gay end of the scale seen in men is somewhat flattened in women. This fluidity is important to recognize and is generally understudied.

While a strict gay-versus-straight dichotomy is not true for either sex, we still have an evolutionary mystery: if there is only one adaptive optimum that human behavioral development is selected to produce (i.e., heterosexual orientation), then exclusive homosexuality, being furthest from it, should be the least common orientation. Because it is not, we can tentatively conclude it is "a thing," a state that needs explanation. While the social conformity model doesn't actually specify a precise distribution of sexual orientations, it offers no explanation here.

A second weakness of the social conformity model is that natural selection is not just about having nonzero offspring—the quantity matters. Women are only able to conceive in a small window each month, and (unlike for other primates) that window is not obvious to either sex. An aversion to straight sex would therefore be expected to cause more monthly ovulation cycles that don't lead to conception. This would extend the average time between births, which in a population would lead to a decrease in family size, even without birth control. Over many generations, a genetic makeup that reliably suppressed same-sex orientation should come to dominate our gene pool. Nevertheless, it has not.

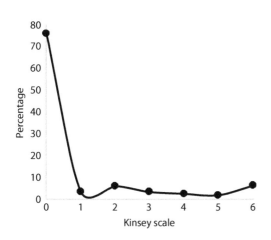

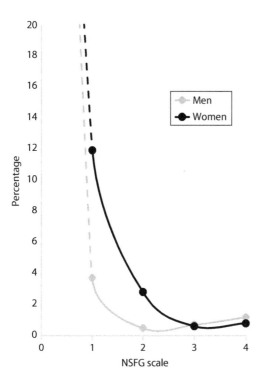

13.2 Distributions of sexual orientation. Top: For men, as measured by Alfred Kinsey in 1949 using a 0–6 scale. Bottom: For men and women, as measured by the Centers for Disease Control and Prevention's (CDC) National Survey of Family Growth (NSFG) using a 0–4 scale. In both scales, 0 represents exclusively heterosexual orientation, while the highest number represents exclusively homosexual orientation. Kinsey generalized his data for white American men ages 16–55 by taking into account the age, marital status, and education of his sample. The CDC data were taken from over 100,000 American men and women ages 16–44. Note how women are much more likely to score themselves as 1 ("mostly heterosexual") and 2 ("bisexual") than men, and how there are nearly as many women at 3 ("mostly homosexual") as 4.

Source: Kinsey data from A. C. Kinsey, W. B. Pomeroy, and C. E. Martin, *Sexual Behavior in the Human Male* (Philadelphia: Saunders, 1948); CDC data from A. Chandra, W. D. Mosher, C. Copen, and C. Sionean, "Sexual Behavior, Sexual Attraction, and Sexual Identity in the United States: Data from the 2006–2008 National Survey of Family Growth," *National Health Statistics Reports*, no. 36 (March 3, 2011): 1–36, https://www.ncbi.nlm.nih.gov/pubmed/21560887.

Option 2: Alternative Reproductive Tactic

In nonhuman animals, males that adopt ARTs are often making the most of a challenging situation. For example, hornless male dung beetles of the species *Ontophagus taurus* consistently develop from larvae that are relatively small.[11] Because success as a horned territorial male requires being large, hornlessness evolved in small males because, at that size, only the alternative sneaker strategy has any chance of success. Almost no beetles have intermediate horns, presumably because this form fails at both strategies. Both the bimodal distribution of human sexual orientation and the substantial reproduction of gay parents are consistent with homosexuality as an ART (or suite of ARTs). The hypothesis here is that being different presents options for reproduction that being typical does not.

It is said that gay male celebrities often have a fan base of adoring straight women. Straight women lament that "all the attractive men are either married or gay." What if we took these cultural memes seriously for a minute? Being different may create opportunities for short-term heterosexual relationships between fundamentally gay men and straight women. Similarly, a butch lesbian may be very appealing to a man with whom she shares male-typical interests and attitudes. More generally, if gay people have traits that make them particularly attractive to a subset of the opposite sex, then substantial fertility may result in spite of their predominately same-sex orientation. It has even been proposed that it is precisely the aloofness of gay people toward the opposite sex that makes them attractive to some members of that opposite sex.[12] We might call this the Liberace hypothesis.[13]

We cannot take a time machine to observe how sexual orientation manifested and had an impact on parenthood rates in ancient societies. The best we can do is examine traditional societies around the world, with commonalities across them likely representing features of our ancestors as they spread around the globe.[14] Paul Vasey, an anthropologist at the University of Lethbridge, and Doug Vander-Laan, a psychologist at the University of Toronto, have reviewed what is known about queerness in men and draw two main conclusions.[15] First, males around the world who are attracted to other males (male *androphiles*) are found to adopt both masculine (*sex-gender congruent*) and feminine (transgender) personas. The latter include the Samoan *fa'afafine*, native North American "two spirit people," and the *bantut* of the southern Philippines. Second, despite their different adult manifestations, both feminine and masculine androphilic men share a similar suite of female-typical childhood behaviors, such as nurturing play with dolls, avoidance of rough-and-tumble play, and increased anxiety when separated from parents. Because of these similar starting points, Vasey and VanderLaan propose that the two types are fundamentally related but diverge under cultural influences.

The early development of a suite of behaviors is consistent with male androphilia representing an ART. But to be an ART, it must also serve as a reliable route to direct paternity. Does it? Vasey has worked for years to characterize the roles of Samoan *fa'afafine* in their society. When I asked about the phenomenon of gay parenthood and whether it might be an important evolutionary factor in its maintenance, Vasey was doubtful: "Homosexual men don't reproduce at all unless there is some external social pressure to do so, usually homophobic. I did a study of 235 *fa'afafine*, and found zero had any reproductive output whatsoever. So, absence of reproduction is not just a Western phenomenon characterizing gay men."

Samoa is not unique in maintaining a stable community of gender-nonconforming androphilic males, but an equivalent tradition is not seen everywhere, and male androphilia seems to be less connected to transgender manifestation in the West. This variety in contemporary human societies clouds our view somewhat, but the evidence we do have suggests that gay parenthood was less common than for heterosexuals in the distant past, probably much less. This puts a damper on the ART hypothesis, and the mystery thus remains: why would a trait that reduces fertility, in some cases to zero, be so common?

Option 3: Unintended Consequence

If homosexuality historically reduced the rate of parenthood, then we are back to the question of how a trait that reduces reproductive success could evolve. One family of explanations proposes that reduced direct reproduction by people with same-sex orientation evolved in the context of family groups and is wrapped up in their overall success. If the benefits of a genetic makeup that occasionally produces gay individuals exceeds the costs in direct fertility that are borne by those individuals, it may persist because of kin selection. Within this kin selection framework, the unintended consequence hypothesis proposes that the same genetic and physiological makeup that enables the occasional development of homosexuality persists because it somehow increases the reproductive success of straight relatives.

One can dream up scenarios for how traits that predispose one for homosexuality could benefit straight relatives. One might be a direct fertility boost. For example, a sort of hyperfeminizing variant could exist that assists women in conceiving or bearing children but predisposes men bearing the same variants to homosexuality. This is not completely far-fetched because close female relatives of gay men do tend to have more children than those of straight controls.[16] We have no idea whether this effect is mediated by reproductive physiology, but it remains possible. The effect may also represent a kind of conscious or unconscious compensation to make up for the childless family member.

Twins are a powerful way to examine the extent to which genetic variation contributes to who has a particular trait. The key is to compare identical and fraternal twins because both share a common womb and home environment, but they differ in their degree of genetic relatedness. Identical twins are, well, identical, while fraternal twins share only half of their genetic variants (as with any siblings). Do these types of twins differ in how matched they are for sexual orientation? Several such twin studies have been done, including one on a set of nearly five thousand Australians led by Northwestern University's Michael Bailey.[17] Bailey and his colleagues found that when they define "gay" rather leniently (a Kinsey score of 1 or more), a gay person's identical twin is 38 percent likely to be gay themselves for men, and 30 percent likely for women. A stricter definition (Kinsey score equal to or greater than 2) reduces the concordances to 20 percent for men and 24 percent for women.

The numbers for identical twins are nowhere near 100 percent, yet they are higher than we expect from the overall population. This suggests there may be a small genetic component to sexual orientation. Still, twins share the same child-rearing environment—could that be the cause? This is where the fraternal twins come in. The concordances for male fraternal twins fall to 6 percent when "gay" is defined as Kinsey greater than or equal to 1 and essentially zero for Kinsey greater than or equal to 2. This striking drop relative to identical twins is a strong indication that genes do play a modest role in determining sexual orientation for men. However, because the identical twin of a gay man is still more likely to be straight than gay, it appears there is no genetic constitution (or *genotype*) that reliably produces a gay individual. This weak genetic effect has been confirmed by association studies that sampled hundreds of thousands of people.[18]

For women, the contribution of genetics is less clear. A fraternal twin of a gay woman is just as likely to be gay as the identical twin of a gay one (30 percent) when the lenient (Kinsey greater than or equal to 1) definition is used. When the stricter definition (Kinsey greater than or equal to 2) was used, fraternal twin concordance was lower, but it was not enough to be distinguished from noise by statistical tests. This difference between men and women likely relates to the different underlying distributions of orientation for men and women discussed above. The twins in Bailey's study closely mirror the CDC data in figure 13.2 in that bisexual orientation (Kinsey scores of 2 to 4) is more common than exclusively gay (Kinsey score equal to 6) for women, but not for men. In addition, twice as many women (5 percent) as men (2.5 percent) rated themselves a 1 (mostly straight but with a slight interest in same-sex partners). As Bailey and his colleagues noted, "Something about male sexual orientation development makes extreme departures from heterosexual development especially likely compared with female sexual orientation development." That "something" for men has a genetic component, but it is also heavily influenced by chance and the environment.

How might genetic variation influence sexual orientation under the unintended consequence hypothesis? It might go something like this: imagine there are one hundred different genes that affect sexual orientation and that each has both straight and queer variants (called *alleles*). If individual reproduction were the only quantity being optimized by evolution, then we might expect everyone to end up with one hundred straight alleles. But perhaps someone bearing ninety straight alleles and ten queer ones is behaviorally straight but more attractive to the opposite sex than the all-straight-alleles alternative. Indulging stereotypes a bit, these might be those sought-after men who are unusually empathetic and dance wonderfully or the women who enjoy rebuilding the transmissions of muscle cars. They are appreciated by the opposite sex, and because they are squarely heterosexual, they have plenty of children. However, the genetic roulette of meiosis could occasionally produce someone with more than ten queer alleles. If a threshold were crossed, they may very likely be gay. This reduces the chance of children considerably for them, but because this more extreme dose is relatively rare and the benefit of lower doses to straight relatives is large, the variants persist in the population. This dynamic predicts that the most exclusive forms of homosexuality should be the rarest. This is inconsistent with the data of Kinsey and others, but that alone cannot rule it out completely.

Another prediction of the unintended consequence model is that the specific attributes of the affected gay family member are immaterial. Instead, it proposes that something occurs in development that creates a mismatch between behavior and anatomy. The evolutionary history of mammalian reproduction may have introduced new opportunities for such mismatches. For example, the evolution of pregnancy from our egg-laying ancestors created, for the first time, a situation where male fetuses are exposed to female sex hormones as they develop. Conversely, mothers are exposed to male-specific proteins during gestation that could affect subsequent male or female fetuses. Under such challenging conditions, perhaps a 97 percent match between sexual anatomy and sexual behavior is as good as can be expected.

Three fellow evolutionary biologists, William Rice, Urban Friberg, and Sergei Gavrilets, recently proposed how unintended consequence might operate (although they didn't call it that).[19] They suggested that reversible modifications to the DNA and associated proteins (*epigenetic marks*) have evolved to reinforce the sexual fate of the parent. This is generally beneficial, but if a parent's marks are retained in any opposite-sex progeny they produce, discordant sexual signals (one from the sex chromosomes and the other from epigenetic marks) lead to a mixture of typical and sex-reversed traits. For example, retained male-promoting marks from the father's sperm might result in female embryos expressing male attributes. In this way, lesbian sexual orientation may develop. Similarly, persistent female-promoting marks in the egg can induce androphilic orientation in a son but have no apparent impact on daughters. If the benefits of these marks for reproductive success are substantial, then an occasional gay offspring is not too large of an evolutionary price to pay. In theory, this could actually work.

The Rice, Friberg, and Gavrilets model is appealing for its ability to explain the substantial yet incomplete concordance of identical twin sexual orientation and for providing a plausible path by which environmental factors could influence development independent of the genome's DNA sequence. Some sort of epigenetic mechanism certainly must be operating because we know than neither genome nor environment are individually sufficient to explain orientation. From a humanitarian perspective, epigenetic reprogramming failure is clearly preferable to maintaining that gay people are intrinsically immoral deviants who have chosen an unnatural lifestyle. While viewing homosexuality as a sort of inadvertent behavioral mistake may increase acceptance, it still tends to reinforce othering. We should therefore accept such an explanation only if we have run out of viable alternatives.

Another way to evaluate the unintended consequence conjecture is to compare the prevalence of traits that we all agree are harmful with that of those that clearly part of normal, healthy human variation. Three well-known congenital disabilities are cystic fibrosis, hemophilia, and spina bifida. They occur in the United States at a rate of one in a thousand to one in five thousand live births. Congenital variants of the genitalia, which provide an anatomical readout of sexual development, are also rare, occurring once in about five thousand births (see chapter 12). Why should the processes that distinguish male and female bodies be so much more consistently executed than sexual orientation? Gay people are not less intrinsically fertile than straights,[20] so they don't have cryptic abnormalities related to the sex organs. There is also no reason (thus far) to believe that the brain would be unusually prone to developmental "mistakes."

In contrast to clearly pathological abnormalities of human development, people also exhibit other minority forms that we don't regard as problematic. For example, red hair and left-handedness occur at a frequency of 1 percent and 10 percent, respectively. The fact that the incidence of homosexuality is in the same frequency range as these benign polymorphisms suggests it is not being limited by natural selection in the same way that true pathologies are.

Like the social conformity model, the unintended consequence hypothesis proposes that exclusively gay orientation is not an alternative morph but rather emerges as the extreme tail of a distribution. Because it requires the simultaneous occurrence of multiple rare events, this tail should be the least populated category, but for men, it is clearly not. Women have a distinct distribution of sexual orientations, but even so there are consistently more exclusive lesbians than we would expect under this model. To the extent to which exclusive homosexuality is overrepresented in each sex, the unintended consequence hypothesis has difficulty explaining it.

The unintended consequence hypothesis also fails to explain a well-replicated finding that relates birth order in families and sexual orientation. The more older siblings of the same sex a child has, the more likely that child is to be gay, and this effect can explain about 30 percent of men with same-sex orientation.[21] The in utero experience of a fetus and/or a young child's family environment therefore can have an impact on sexual orientation in a way that is influenced by earlier arrivals. How this works is a mystery,[22] but failure of epigenetic reset or other, purely genetic versions of the unintended consequence hypothesis cannot easily explain it.

Another reason to remain skeptical of the unintended consequence hypothesis stems from the often tacit view that homosexuality represents an unselected, fundamentally pathological state. My complaint here is admittedly more anecdotal yet hard to deny: people with homosexual orientation just don't seem to be sickly. They are well represented among the very most accomplished scientists, writers, business people, artists, and athletes that the world has known. From Billy Strayhorn to Billie Jean King, from to Alan Turing to Tim Cook, there is a clear sense that one's sexual orientation has no bearing on one's potential to thrive in the most demanding of human endeavors. What is being missed here?

Option 4: Social Reproduction

We have seen that alternative sexual morphs are widespread in the animal world and that in some cases kin selection promotes their evolution. In social species, many individuals do not reproduce at all. These nonreproductive individuals are clearly not random developmental noise but instead emerge early in life as a distinct form. The similarly early manifestations of queerness present a striking parallel with these animal examples. Just as cryptic male salmon diverge from standard-issue male development early in life to adopt the alternative form, many boys and girls who will be queer adults can be recognized in childhood by developmental psychologists and unfortunately by playground bullies.[23] It thus appears that, for at least a subset of gay people, a suite of correlated traits develops alongside adult sexual orientation. These traits may be at the root of gay stereotypes. Seen from the polyphenism perspective, however, there is no reason not to acknowledge their existence as long as doing so can be decoupled from discrimination and homophobia.

The following question remains, however: what would such a polyphenism achieve? The ART scenario discussed above is one possibility, but the evidence that gay people reproduce at much lower rates across multiple cultures suggests we need an alternative. The answer may lie in kin selection. We are an intensely social species with no biological weapons other than our large brains and opposable thumbs. We are vulnerable as individuals, and especially as children, yet mighty when working together. As our unique language capacities show, group functioning has been a higher-order target of natural selection in our species for millions of years.[24] Behavior that increases the survival and success of one's siblings, nieces, or nephews (and of their other caregivers) can also be adaptive. This brings us to the social reproduction variant of kin selection. In this scenario, homosexuality is a concerted strategy (or set of strategies) that evolved under natural selection to enhance group success. As a result, the specific attributes of gay group members are actually adaptive rather than reflecting an incoherent suite of "mistakes" that is paid for by a direct fertility boost to straight relatives.

This social reproduction hypothesis predicts that we should expect to see gay family members somehow supporting their straight kin in having kids. For example, they might be unusually inclined to (or effective in) helping their young relatives. Is there any evidence for such uncle-like behavior (called *avuncularity*)? Measuring this is tricky, and showing that it has a direct impact on the survival of relatives even more so. Perhaps it is not surprising then that research thus far has been sparse and rather inconclusive. Vasey's ongoing work in Samoa has provided some support for the idea that androphilic men are more inclined to support the children of their close relatives than are straight men.[25] However, attempts to replicate these results in the industrialized West have not been successful.

Even if playing the role of the generous uncle or auntie is not the norm for queer people, we could imagine other ways in which their presence in one's family or village becomes an asset. By contributing to group success economically but not adding to its population growth, they may materially enrich the lives of their relatives. For example, by tending crops, hunting, gathering, or cooking, the food supply per group member may be increased relative to the alternative if that same family member had children of their own. This in turn might reduce infant or childhood mortality enough to more than compensate for the lack of parenthood for the gay person. But for this to be distinct from the unintended consequence model, the overall vigor and specific characteristics of gay people should be important. What these characteristics might be is murky at best. One possibility is that the ability of straight kin to recognize queer orientations reduces group tensions. Parents might feel more comfortable having their adolescent daughter work alongside a gay man than a straight one. Or, if we are like bonobos, sex between females or males may have played an important role in holding the group together.

If homosexuality evolved as a way to create a more favorable adult-to-child ratio, then we might also expect cultural mechanisms that achieve the same result to appear. Anthropologists working in contemporary societies have documented a variety of traditions that have the effect of limiting reproduction of both men and women. For example, Tibetan farmers living in the harsh climate of the Himalayas often designate a daughter to be a nun. After religious training, the young women return to their families to perform a wide range of household-supporting tasks, including caring for younger siblings and aged parents, farming, and tending to livestock.[26] Similarly, the priesthood in Catholic Ireland was disproportionately pursued by men with an unusually large numbers of brothers.[27] These planned celibacy traditions likely arose in response to tight constraints on land availability for subsistence farming, in which oversubdivision of land leads to unsustainable parcels. They also mirror, and even reinforce, the effect of birth order described above, and thus for some provided a socially acceptable closet.

The existence of same-sex orientation in all cultures implies that it evolved in preagricultural human societies. If it did, then other types of resource limitations, such as access to food, water, or natural shelter, may have led to the same need for restrained reproduction. According to Janet Chernela, a cultural anthropologist at the University of Maryland who has worked extensively with Indigenous Amazonians, "there is ample evidence of the need for mobile peoples to limit population size. They live in very small groups, and cannot move with more than one pre-walker. Every effort is made to control reproduction: contraceptive herbs, long periods of abstinence, and even infanticide."

If the management of population growth has been a human concern for a very long time, it is possible that homosexuality evolved via kin selection to allow sustainable growth and decreased mortality. Testing this idea is challenging, but the prediction would be that a gay-friendly culture reaps the benefits of having what are effectively extra parents and workers, while an otherwise similar one that forces gays into the closet (including any unwanted parenthood that might come with it) would suffer higher mortality.

WHO AMONG US?

Kin selection may explain *why* homosexuality evolved, and several of its features suggest it may represent an unrecognized polyphenism, perhaps one that facilitates group survival. If kin selection makes having a subset of less-reproductive members beneficial to a family or group, does it matter *who* occupies that role?

Perhaps not. Gay orientation may develop at random as the result of an epigenetic process that evolved to produce that outcome at some reasonably constant rate. Any more specific model would have to outperform this in terms of predictive power in order to be taken seriously.

Arguing against the at-random model is the strong effect of birth order on sexual orientation. In particular, men with multiple older brothers are more likely to be gay. This is consistent with making the best of a bad situation. Perhaps archaic humans lived in family groups in which not all individuals received adequate resources for successful reproduction (just as some do not today). Early-born kids may have been given (or they may have taken) most of the resources that are crucial for forming successful mate pairs. Under such circumstances, being the youngest of several children of the same sex may be a great disadvantage from the standpoint of mate-finding and direct reproduction. A switch in strategy to avoid direct competition with older same-sex siblings could thus be adaptive. While birth order alone cannot consistently predict sexual orientation, queerness seems to mark disproportionately the most appropriate group members to respond to kin selection.

CONCLUSIONS

We are not salmon or dung beetles, and we need to be careful in making overly facile analogies with animals.[28] Nevertheless, after three decades of research on animal reproduction, I am continually astounded by the combined power of sex and evolution to shape anatomy, behavior, and the genome. Despite earlier objections to the contrary,[29] it is increasingly evident that the innate behaviors of our own species include sexual orientation. All these behaviors were shaped by natural selection, with our intense sociality adding a strong dose of kin selection.

The high frequency and early development of same-sex orientation, as well as the relative rarity of bisexuality in men, all point to the possibility that our own species may have an alternative reproductive form of whose existence we have thus far been unaware. Such a reproduction-suppressing morph would evolve only as an outcome of kin selection. Its trigger by environmental variables makes it analogous to the induced reproductive arrest seen in other social animals. Of the four models discussed in this chapter, these observations fit the social reproduction hypothesis best.

The evidence for a sexual orientation as a polyphenism is stronger for men than for women, and we know very little about how it develops. Nevertheless, recognition of such an alternative state, even if it explains only a subset of situations, would have profound implications for human health and happiness. We no longer try to convert lefties to righties (although we once did). Regardless of the evolutionary dynamics that got us here, it is hard to avoid the conclusion that queerness, while not the *norm*, is nevertheless *normal*. It may even exist for a purpose that has been beneficial for human survival for thousands of years. Clarifying this will require much more research—may we get down to doing it.

CHAPTER 14

Pronouns

Synopsis

The number of openly transgender people in the West has increased dramatically since 2000. Why? Much remains unclear, but there are likely two distinct forces at work. One is a deep-seated sense of one's biological sex being the opposite of one's internal identity, often recognized early in life as gender dysphoria. The other might be termed "gender resistance," a visceral refusal to adopt the assumptions and limits that are culturally associated with our sex. As social and legal space to be gender nonconforming opens, more people in each category are able to transition.

———— ◆ ————

In the early 2010s, a teenager who lived up the street from me made two life-altering decisions. First, they enrolled in a nearby university. Second, having thus far presented as a girl, they announced to their family that they were nonbinary and changed their name and appearance to underscore that. Having seen them grow up as a spirited but outwardly feminine kid, I didn't see it coming. A couple of years later, another couple we know informed us that their preschooler, who was born male, was expressing a strong female identity. They wanted to support their child, and soon clothing and pronouns were changed to convey that they actually had a daughter. In the last few years, many other friends have shared similar news.

Most people living in twenty-first-century North America can tell stories like these, but I don't recall this from my own childhood in the 1970s and 1980s. I lived in a small, heavily Mormon Nevada town, so perhaps it isn't surprising.

But once I got my driver's license, I spent as much time as possible over the pass in the Las Vegas valley, with a shifting circle of artsy friends in a town in which sexual freedom was a cherished virtue. If there were a safe place to be an out trans youth in 1980s Nevada, this was it, but I don't recall anyone fitting the contemporary definition.

In college, I met gay rights activists driven by the urgency of the AIDS crisis. They demanded that straight people acknowledge their humanity and their suffering and that they join them in insisting that government do the same. This gay activism was highly effective yet decidedly cisgender. My music classes introduced me to electronic music pioneer Wendy Carlos (a trans woman whose *Switched on Bach* recording was a commercial success), but upon graduation in 1990, I had yet to meet an out trans person myself. That would not occur until the mid-1990s, when a trans woman joined the lab in which I was completing my PhD research.

Historian Susan Stryker helped me understand that my experience was rather typical for a straight, cisgender guy of my age.[1] In the 1980s, trans activism was beginning, but people openly living out their trans-identity were relatively rare, and they were concentrated in a few cities. Their message was blunted further by a cold shoulder from others who were also struggling for gender-related justice. In particular, some prominent feminists were unwilling to accept trans women as fellow women or even as comrades in the struggle. Among the most strident academic opponents of trans normalization has been Janice Raymond, who spent most of her career at the University of Massachusetts. Her 1979 book *The Transsexual Empire: The Making of the She-Male* attacks the motives of trans women, arguing that they practiced a sort of stealth misogyny. In the foreword to the book's reissue fifteen years later, she asserts "transsexualism constitutes a sociopolitical program that is undercutting the movement to eradicate sex-role stereotyping and oppression."

Between the expected opposition from social conservatives and rejection from so-called trans-exclusionary radical feminists (TERFs), advocates for trans rights had (and still have) a steep hill to climb. Yet slowly they made headway. Trans man Lou Sullivan founded FTM ("female to male") International in San Francisco in 1986, broadening the discussion from the initial emphasis on trans women. The 1989 U.S. Supreme Court decision in *Price Waterhouse v. Hopkins* found that a major accounting firm engaged in illegal discrimination when it denied partnership to Ann Hopkins, a ciswoman and mother of three, for being too masculine. Perhaps more important, Raymond's conspiratorial critique came to be weighed against the increasing media coverage of the lived experiences of many trans women, especially those of color. Rather than sitting atop a group of subservient ciswomen, many were instead without housing, struggling with substance abuse, ostracized by their families, and being routinely murdered on the streets while engaged in risky sex work. Even those who cannot regard trans women and ciswomen as "the same" began to see that this was an academic dispute compared to the pressing need for basic human rights.

The last twenty years have seen a surge of awareness of trans-identities in the United States and other Western countries. A stream of books, web and print articles, award-winning TV shows like *Pose*, and even an entire 2017 issue of *National Geographic* have been dedicated to the variety of gender identities. A significant chunk of our daily news centers on the advances of transgender people and the cultural and political backlash against them. Is this largely a media phenomenon or has there really been a change? More precisely, is mainstream straight, cisgender culture just finally acknowledging the existence of a fairly stable population that was always there or has there been a fundamental demographic shift? Both could be true to some extent, and only recently have the sort of data required to address the relative impacts of core identity and a more inclusive society become available.

THE NUMBERS

Measuring the size of the trans population comes with two main challenges. First, until fairly recently, nobody was doing the measuring. While irrefutable evidence of transpeople in the West goes back centuries, it is anecdotal (that is, not quantitative). Broad population surveys of the kind routinely used in epidemiology were unheard of before 1960. They only became common enough to discern trends after 1980. Second, the yardsticks used to do the measuring are imperfect. They have not been standardized around the world, and any specific instrument will have inherent limitations. For example, transpeople who are not "out" will avoid transitioning to an extent that depends greatly on the part of the world where they live. Similarly, asking people about their identity opens a range of interpretations that may be mixing different situations together. Patchy as it may be, however, the data available in 2023 can already paint a low-resolution picture.

Figure 14.1 shows data from a large set of published studies on trans-identity, compiled by an international team led by Emory University's Michael Goodman (with a few of my own updates).[2] Most are from Europe and North America, with a few cases from East Asia. Its three measures of trans frequency come from clinician reports (for people seeking medical transition) and, more recently, from broad surveys of the general public. From these we can draw several conclusions:

- There has been a sharp increase in the frequency of gender dysphoria diagnoses in the West. From the perspective of psychiatry, these are people that experience distress stemming from their sense of self not matching their sex assigned at birth. Not all people with gender dysphoria are actively undergoing gender-affirming clinical interventions (including hormone treatments and surgery), but those who are can be counted. While studies conducted by different teams give different numbers (an "apples to pears" comparison that introduces variation), those repeated by the same team on the same population ("apples to apples") also show a striking upward trend over time. Current estimates from clinical activity are that ten to fifty people per 100,000 (that is, 0.01 to 0.05 percent) are gender dysphoric.[3]
- The frequency of those identifying as transgender (independent of transition-related medical care) is vastly higher. The current incidence is roughly 1,000 per 100,000, or 1 percent.
- Where the data are collected matters greatly. The three Asian studies (one in each panel) yielded estimates that were very high compared to similar estimates in the West at that time. This likely reflects the existence of long-standing third-gender traditions in Asian and Pacific Island cultures, such as the *fa-afafine* of Samoa (see chapter 13) and the *kathoey* of Southeast Asia.[4] More recently, however, Western incidence has largely caught up.

The increase in the amount of gender-affirming care for gender dysphoria indicates that many people suppressed a core trans-identity in the past. Given the many risks of coming out trans, this is not surprising. But how big then is the trans population? Because an unknown fraction of transpeople remain out of contact with specialized health care, this is difficult to say with precision. What we can say, however, is that people currently engaging their doctors about

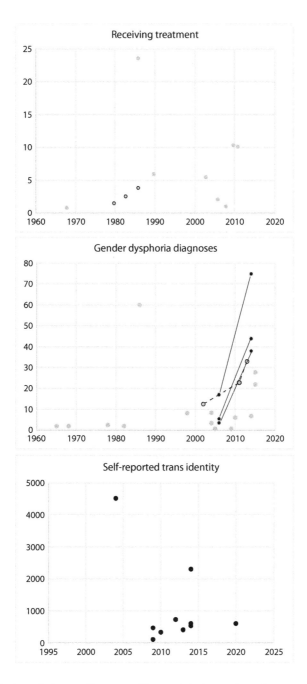

14.1 Historical trends in trans frequency. The *y*-axis is the number of cases per 100,000 total population. *Top*: Data from surveys of providers of gender-affirming hormone treatment or surgery. The three darker points are from the same set of Dutch clinics. *Middle*: Diagnoses of gender dysphoria. Longitudinal data from the same study population are connected by lines. *Bottom*: Self-reported trans-identity from broad population surveys. High outliers in 1986 (*top, middle*) and 2004 (*bottom*) are from Singapore (*top*) and Taiwan (*middle and bottom*). These are the sole Asian samples in each case.

transition are still less than one-tenth as common as people who identify as transgender. This is noteworthy and deserves some consideration.

It seems highly unlikely that the entire population of out transpeople who never undergo hormone therapy or gender-affirming surgery really want to but do not or cannot. Are there other drivers of trans-identity than gender dysphoria? If so, it's not hard to imagine what they might be: In a deeply gendered world, the norms associated with one's natal sex will often chafe. Those born female who don't feel consistently or at all feminine and males who don't feel masculine are nevertheless continually expected to affirm canonical cisgender behaviors and attitudes, as long as they look the part. Refusing to look or act the part, in the gender-as-performance sense,[5] offers a way to resist this. This suggests a hypothesis: a large fraction of transpeople who forgo medically assisted transition may not be experiencing gender dysphoria as defined by mental health professionals (i.e., the need to be the other sex in a binary conception) but rather are manifesting a second cause of trans-identity that we might call term "gender resistance."

One thing that both gender dysphoria and gender resistance have in common is the rejection of the ways in which being cis doesn't fit or serve. From this perspective, they can be thought of as "pushes" that move people's gender identity away from cis. This is a rather negative conception, however. Are there also "pulls," or positive attractors, at play? A growing literature of trans scholarship indicates there are.[6] Finding joy in gendered aspects of our lives is something that many cis-people take for granted, but it is often true for transpeople as well when they transition. This pull to an identity that truly fits has been dubbed "gender euphoria."[7]

TESTING THE GENDER RESISTANCE HYPOTHESIS

If gender resistance underlies a large fraction of trans-identities, then we can make some predictions. First, because it forms as an interaction between innate disposition and social pressures, we might expect it often to have a later onset than canonical gender dysphoria. Second, we predict that rejection of one's default cisgender could lead to a range of alternatives, from adopting the other binary gender to embracing a nonbinary or agender identity. Thus, there should be a large and stable fraction of transpeople whose gender identity is not clearly male or female. Finally, we expect that, for many, their final transition goals stop short of surgeries.

In 2016, the National Center for Transgender Equality (NCTE) released results from a nationwide survey of 27,715 transpeople.[8] This was the biggest dataset of its kind to that point, and several of the survey's questions address the development and variety of trans-identity. When survey respondents were asked about when they began to feel they were transgender, 10 percent reported it was before the age of five, and another 16 percent before the age of ten. This population likely contains most cases of "traditional" gender dysphoria. A substantially larger proportion (28 percent) reported reaching this milestone between the ages of eleven and fifteen (i.e. during puberty), 29 percent at ages sixteen to twenty, and another 10 percent at ages twenty-one to twenty-five. These numbers for later onset are consistent with gender resistance as a driver of trans-identity.

Now things get tricky. If the social component of nondysphoric trans-identity is a key factor in its emergence, does that mean it is "contagious"? Many conservatives are terrified of this exact possibility, so we need to tread carefully. Even in academic circles, suggestions of peer influence in the emergence of adolescent transgender identity has been highly controversial. Most notably, an article describing apparent cases of rapid onset gender dysphoria (ROGD)[9] led to a flurry of rebuttals, which criticized it for its flawed survey methodology.[10] The leading professional societies in psychiatry and clinical psychology have also made clear that ROGD is not a diagnosable mental health condition.[11]

Although ROGD may have been discredited, there really are some teens who declare themselves trans with little outward warning to their family and friends. What are we to make of that? Some are certainly cases of a trans-identity (with associated gender dysphoria) that was undiagnosed and suppressed for years before finally being revealed. However, their existence is expected under the gender resistance model and should not be lumped into the same category as persistent, early-onset gender dysphoria. Gender dysphoria (but not gender nonconformity per se) continues to be considered a diagnosable mental disorder.[12] As conceived of here, gender resistance cannot be regarded similarly—it is a rational, principled reaction to our highly gendered world.

With regard to the current prevalence of nonbinary identities, the NCTE survey found that 31 percent of respondents were comfortable with this term being used to describe them. "Genderqueer," "gender nonconforming," "genderfluid," and "androgynous" were acceptable to 18 to 29 percent. Equally informative was the finding that, while 65 percent of respondents would describe themselves as "trans," only 18 percent would describe themselves as "transsexual." This indicates there is indeed a large fraction of transpeople for whom hormonal and surgical transition is not the end goal. Reinforcing this impression, 82 percent of trans men and 68 percent of trans women say that they have completed their transitions, which is a much larger fraction than the numbers for clinical treatment would indicate. Consistent with this, a 2022 Kaiser Family Foundation (KFF)/*Washington Post* survey of over five hundred trans Americans found that two-thirds prefer to use them/they pronouns some or all the time. Passing as the other binary sex is often not the point.[13] For many, it seems that changes in clothing, hair, pronouns, and identity documents suffice.

I discussed some of this chapter's themes with a cisgendered female friend who grew up in Latin America in the 1970s. Living now in the United States, she is comfortably feminine and a mother yet also sympathetic to gender nonconformity. She recounted how, as a girl, she envied the freedom that boys had. Freedom to be rough, freedom to get dirty, freedom to roam without as much supervision, freedoms she did not have because of the norms of female behavior that were rigorously enforced by the adults in her life. And there have been countless men who didn't feel the norms of boyhood or manhood fit them but didn't feel a strong need to be women. "I think gender is this generation's fight, and I get it," she said. This is the crux of the gender resistance idea: in a free society, we should be able to choose to say no to societal expectations associated with our biological sex, and it is our fundamental human right to do so.

As this book goes to press, an anti-trans political backlash is sweeping parts of the United States. It began with campaigns to block hormonal or surgical interventions for minors. There is a legitimate ethical issue here (see below), and it resonated with many. What was rarely pointed out, however, was that this was a sensationalist red herring: surgeries have never been part of the standard of care for trans children, nor do trans activists promote them. Many transpeople and their allies smelled a rat and rightly so. Conservative politicians are generally happy to leave medical decisions related to treatment of nonsexual conditions, such as heart malformations or pediatric cancer, to the patients, parents, and doctors closest to the situation. As with kids suffering from these conditions, trans youth are also at risk of dying (primarily from suicide). This suggests that the urge to regulate the complex intersection of gender identity and mental health is not just about protecting kids from decisions they might some day regret. Rather, there is something about questioning of gender per se that is threatening to these politicians and/or their constituents. Consistent with this, several states are currently considering bills that would block gender-affirming interventions for trans adults as well.

CROSSING OVER

When someone seeks medical care to address gender dysphoria, the only real option at present is to bring the body in line with the internal identity. On the less drastic end are cross-sexed hormone regimens, which can achieve substantial change in outward appearance even in adults. For example, transfeminine people born male can grow breasts, and those born female can grow facial hair and even develop male pattern baldness. However, as described in chapter 2, many developmental differences occur before birth, and these are generally irreversible through hormones alone. Some changes that develop during puberty are also irreversible without surgery, such as facial contours, shoulder width, vocal pitch, and the Adam's apple.

The sexual biology of our species thus presents a dilemma: the earlier a transperson begins to receive gender-affirming interventions, such as hormones or surgeries, the more likely they are to "pass." At the same time, our legal systems generally regard people younger than eighteen as too young to make their own health-care decisions, at least independent of their parents or guardians. It is no surprise therefore that the age at which gender-related care is allowed is a focal topic of social and legal debate.

The irreversibility of much of human sexual development leads to stories like that of Freddy McConnell. McConnell was female at birth but began to take testosterone at age twenty-five and at twenty-six had top surgery to remove his breasts. He presented as a man (mustache, beard, etc.), but because he retained a functional female reproductive tract, he retained the option to become pregnant. In 2019, he acted on this potential and delivered a healthy baby.[14]

For the subset of transpeople who seek bottom surgery, advances have certainly been made. Nevertheless, this is a major reconstruction, and some important aspects of normal function cannot be achieved. With regard to sexual pleasure, a vagina constructed from a male starting point will lack the glands that normally lubricate it during sex, and a penis built from a female precursor will be incapable of a normal erection without some sort of prosthesis. With regard to reproduction, the reconfiguration of the genitalia comes at the price of completely sterilizing the patient. That most people who have undergone bottom surgeries nevertheless report big improvements in their mental health after transitioning is an indication of just how debilitating gender dysphoria (and/or how strong the pull toward gender euphoria) can be.[15]

It is at least worth asking whether gender dysphoria could or ever should be addressed by altering the internal gender identity instead. This is admittedly an odd thing to contemplate because our gender identity is a major part of our sense of self. If it were changed, would I still be "me"? It is also a touchy subject because it could be an opening for justifying involuntary "treatment" of transpeople to make them conform. It cannot be denied, however, that if such a treatment were actually available and effective, it would alleviate dysphoria while sparing some people arduous interventions. For the foreseeable future, however, there is little hope (or risk?) of such an alternative. Although we understand the ways in which male and female fruit fly brains differ down to the level of individual genes and cells,[16] efforts to do the same for humans have made only modest progress.

That we don't yet know what in our brains makes us feel male, female, both, or something else doesn't make the feeling any less real. Somewhere in all those billions of neurons in our brains are the circuits that encode our gender identity. While these circuits may be subject to environmental influence, there must be an innate starting point strongly influenced by our biological sex. For example, a large majority of people are innately sexually attracted to the opposite biological sex (instead of it being a 50–50 coin toss), and males and females differ greatly in their incidence of disorders like autism (more common in males) and depression (more common in females).

Neuroscience is still far from being able to provide a clear model for how sexual orientation and gender identity are wired into our brains, but some progress has been made. In 1991, the British American neurobiologist Simon LeVay compared a brain region known to be central to sexual behavior in monkeys (the anterior hypothalamus, specifically, the INAH3 nucleus) in samples of straight cismen, straight ciswomen, and gay cismen.[17] LeVay confirmed previous findings of a sex tendency in the size of the INAH3, but he also found that this region in gay men was, on average, more like that of straight women. This work required portmortem dissection of brains, material that was sadly abundant at this time as young, gay men died in appalling numbers from HIV-AIDS in the 1980s.

LeVay's article was controversial at the time, but it helped legitimize a search for how our brains encode sexual orientation and gender identity. In the three decades since, its central finding has been replicated and extended to gender identity.[18] Dutch neuroscientist Dick Swaab and his colleagues have reported that INAH3 size also differs on average between cisgendered and transgender people who were male at birth.[19] Using less invasive magnetic resonance imaging (MRI) brain imaging on live subjects, a number of studies have found aspects of brain anatomy and activity that correlate with gender identity.[20] Changes in sex hormone profiles also affect the brain activity of transpeople as they transition.[21] This is all consistent with the idea that gender identity is "real" inasmuch as it is associated with measurable anatomical and physiological states. So much remains elusive, however, and we are hampered by the more general inscrutability of the vastly complex human brain.

Eventually the physiological embodiment of gender identity may be revealed. If and when that day comes, the questions will then become whether changes in that neural system can be affected in a way that is any more artful than the gender-affirming interventions available today, and if so whether they *should* be. It is quite premature to make any predictions here. One thing that seems clear, however, is that, for the majority of transpeople not considering surgery, their maximal well-being will likely come less from advances in medicine and more from the "treatments" that need to be administered to society.

Epilogue

Eyes Wide Open

Synopsis

Sex and reproduction have permeated our anatomy, our behavior, and our health since long before we were humans. Efforts to create justice that start with the premise that human sexes are fundamentally a cultural construct are therefore doomed. At the same time, modernity now allows our species to decouple biological sex from gender and from power. We are a species in the midst of a massive transition, one marked by tension between historical inertia and new possibilities.

———•———

Settling into the steaming hot bath in Kyoto's Kurama Onsen, I was oddly alone in a country where privacy is not easy to come by. The near-scalding water felt great, and the view out of the floor-to-ceiling windows was breathtaking. Marching up the hill outside, enormous silver-barked *sugi* trees formed an open forest unlike anything I have seen in North America. Happily cooking in the bubbles, I began to notice other things about my surroundings. The baths, as everywhere in Japan, are separated into male and female sides. This is mostly because everyone is clad only in their birthday suit, except for a hand towel most people place on the top of their heads. I could hear my wife and daughters enjoying themselves on the other side of the partition that divided us. As I appreciated the view, I wondered if they were doing the same. And then I realized they couldn't: there was a tall wall blocking their view that the men's side lacked. And then I wondered why.

Creeps. Yes, probably creeps. While someone might try to sneak down the wooded slope to a point where they could check out the men and boys, the risk of a peeping tom harassing the women was deemed serious enough that it had to be prevented at considerable expense, both in terms of masonry and in depriving the females of the view I freely enjoyed. I am not an expert on Japanese voyeurism, but I know it happens. Unfortunately, it happens everywhere. Every week or so I get an email from my university's public safety office that is similar to this one:

> On October 25, 2022, at approximately 1:02 p.m., the University of Maryland Police Department responded along with the Prince George's County Police Department for an off-campus indecent exposure that had just occurred. A woman reported to police that she was walking along Yale Avenue, past Lehigh Road, when she saw a dark blue sedan with its window down. Inside the vehicle was a man who exposed himself. The Prince George's County Police Department is investigating this incident.

Note that it is *never* a woman exposing herself to a man. How much of male creepiness depends upon innate biology, and how much of it is learned and/or facilitated by cultural inertia? What about all the other ways in which the existence of male and female sexes connect to inequality, anxiety, and suffering? Might anything we have seen in the last fourteen chapters help us find winning strategies among the many proposed? I believe it could.

A SPECIES IN TRANSITION

In addressing the many societal ills with sexual roots, the basic situation is historically entrained and quite messy. It should now be clear that we cannot dismiss sexes as purely social constructs or arbitrary divisions of a seamless continuum. The roots of male and female in our species go back hundreds of millions of years. To recap: after about one billion years of sex as a unicellular phenomenon, overt sex difference (anisogamy) evolved in the first animals. The separation of sperm and eggs into the specialized male and female bodies that we have today was set in our earliest vertebrate ancestors. As some fish moved onto dry land and evolved into amniotes, both internal and external anatomy diverged between the sexes to allow internal fertilization. New opportunities promoted a shift to pregnancy and live birth in mammals. In the primates, sexual activity was disconnected from ovulation and thus from reproduction to a large extent. Standard sexual selection of males in our great ape ancestors led to differences in physical strength in males and females, which remain in a reduced form today. More recently, our elongated childhood promoted an unusual dynamic of mutual mate choice, which has selected for female-specific ornaments. The sex differences of male and female bodies evolved in the context of a visually focused assessment of mates, which emerged as the role of smell was reduced. All of this is who we are.

Throughout this book, I have generally used the commonly accepted convention of distinguishing sex as biology (i.e., anatomy, hormones, development), while gender is socially constructed. Such a distinction is important because it allows space to consider each independently, but it does not preclude connections between them. Over the vast expanse of premodern time during which we became human, just getting enough to eat while raising children was once an all-consuming (and frequently unsuccessful) challenge. As a result, one's biological sex influenced one's social and family roles in ways that were largely nonnegotiable.

What do I mean by nonnegotiable? In a recent seminar class I taught, the students and I considered a hypothetical community of ancient families that routinely shared key resources with each other. Perhaps one day it was division of a mammoth carcass, the next, access to a prime fruit tree or the really good planting spot. There would be times when a woman would be unable to represent her family in the negotiations, such as shortly after childbirth. In addition, not everyone is nice, and in a mixed group of representatives, some males may be tempted to resort to physical intimidation of females to get more than their share. This might promote a "send your man" pattern that would be hard for even a feminist family to buck. Although this is speculative, it is a plausible dynamic that may have set in motion the more public male, more domestic female pattern that characterizes many human societies. There may have also been a mate-guarding component as well.

Much has changed in the developed world. We can buy food, there are legal protections for vulnerable people, contraception is (mostly) available. There are also now many configurations of families that are viable, both materially and emotionally. Nevertheless, the way we are regarded by our fellow humans is still influenced by our sexualized anatomy. It seems unlikely we will ever completely decouple gender from sex. Nor do many of us even *want* to do so. University of Amsterdam philosopher Mari Mikkola captures the tension here nicely:

> First, claiming that gender is socially constructed implies that the existence of women and men is a mind-dependent matter. This suggests that we can do away with women and men simply by altering some social practices, conventions or conditions on which gender depends (whatever those are). However, ordinary social agents find this unintuitive given that (ordinarily) sex and gender are not distinguished. Second, claiming that gender is a product of oppressive social forces suggests that doing away with women and men should be feminism's political goal. But this harbours ontologically undesirable commitments since many ordinary social agents view their gender to be a source of positive value. So, feminism seems to want to do away with something that should not be done away with, which is unlikely to motivate social agents to act in ways that aim at gender justice.[1]

As one such ordinary social agent living in the twenty-first century, I see her point. Yet this suggestion that the sex-gender connection can hold positive value is not synonymous with arguing that sex is or should be destiny. A cartoon of a biodeterministic position is something like this: "Men are sex-crazed and generally stronger than women, and this is so imbedded in our biology and social interactions that we should just accept (heck, even promote) an objectified and subordinate status for females. Plus, because straight people outnumber queer people, we can regard them as abnormal and ignore their rights."

There will always be fans of this view, including males who appreciate the perks of dominance, some females living in conservative cultures who want their choices validated, and those afraid that gender nonconformity is contagious. Although some traditional gender roles may have been necessary in primordial societies to some extent, the options in the developed world today all function to erode the importance of sex differences in our daily lives. This seems like progress, yet it is important to see that the conditions allowing this are very recent and not yet found in all societies. No wonder we are confused.

Today there are countless men who are their kids' primary caregivers or who work as nurses, and similarly innumerable women who serve in the military, drive trucks, and lead corporations. Same-sex relationships are common, and people born as one sex identify as the other or neither and express this. Society has changed as a result, but it has not collapsed. The notion of precious, divinely mandated sex roles still appeals to many, but its rigid enforcement serves no purpose and is the source of much human misery. The freedom to be oneself in a gendered world is an essential human right, but it is clear we are still learning how to both assert our individual identity and how to accept it in others.

VALUE JUDGMENTS

If our priorities can now be largely disconnected from our sex, it is worth considering what is judged as "best." In particular, is the value of the public and external (often associated with males) actually different from the private and domestic (often associated with females)? From the standpoint of monetary pay, the former is certainly more valuable in the developed world. Early feminists like Simone de Beauvoir saw menstruation, pregnancy, and nursing as obstacles women had to overcome to reach equal footing as men. But what about "value" in the more karmic sense? For example, which kind of activity actually influences the next generation more?

Philosopher Georgia Warnke uses the notion of "projects" (i.e., activities visible to the community) to represent the kind of public tasks that men have traditionally emphasized and that are compensated with money in modern life. In the West, most women now have substantially higher levels of education and jobs outside the home, and they enjoy the money, independence, and public life that goes with them. This turn toward projects feels like progress, yet it is also associated with opting not to have children. Nobody should be compelled to be a parent. At the same time, everyone currently kicking butt in medicine, music, mocha making, or management had to have parents. Within limits, having kids must therefore actually be a public good. Might the birthing and rearing of children be a kind of project, too? I certainly know plenty of mothers (including my own) who regard their children as their projects, but they weren't paid for it in cash. Were caregivers of *either* sex to be compensated for raising kids, it might change this dynamic substantially in a way that benefits parents, kids, and society at large.

The notion of public versus familial domains and the patriarchy associated with that fits a Western pattern, but other cultures have done it differently. An important example is the precolonial Yoruban society of West Africa, centered on present-day Nigeria.[2] A distinction between male and female was recognized, both in regard to marriage and reproduction and as it related to practical division of labor. Regarding power, however, the sexes were not perceived as antithetical /oppositional as in the West, nor were there any connotations of dominance or subordination in relation to sex. Thus, Western ideas of the genders "man" and "woman" didn't really exist. Instead, Yoruban social status was traditionally determined by seniority rather than sex. Being senior conveyed both rights and responsibilities. This remarkable system persisted until colonialism imposed Western gender roles.

The Yoruban model was abetted by a language that lacks gender-related features we take for granted in English. Yoruba has no gendered pronouns, no sex-typed names, nor gender-specific relational categories (one word for 'aunt" and "uncle," one word for "son" and "daughter," etc.). The Yoruba term *oba* is often translated in English as "king," but it is actually a gender-neutral term for ruler. While aspects of gender-neutral language are also found in societies that nevertheless have strong gender roles (e.g., China before 1900),[3] it does seem to predispose the mind to seeing past biological sex in evaluating unrelated issues. What if we all went by "they" or a newly coined singular equivalent? Current efforts to promote pronoun diversity in English are thus a worthwhile experiment.

Despite both historical and present-day examples of what a less-gendered world looks like, many are still very uncomfortable with that world. Cultural inertia ("but we've always done it this way") is not a good reason to maintain involuntary, oppressive structures related to sex and gender. At the same time, sex and gender often underlie positive identities, and many of these will be "traditional." This also needs to be respected. A plural approach is needed.

MOVING FORWARD

Assuming you share my desire for a world that is equitable with regard to sex and gender, how do we make progress? As with all struggles for justice, it necessarily will involve informed agitation. It is my hope that some of the content of this book will serve this purpose. Sex-related cultural norms can have ancient roots, yet time and activism can open people's eyes. The rule of law can transform society (at least in places fortunate enough to have it). If we look at the United States over the time I have been alive, an impressive example of the activism-plus-law combination is Title IX of the Education Amendments of 1972 and its impact on sports. By requiring institutions to invest in men's and women's sports equally, a whole generation of women athletes have emerged in sports traditionally barred to them, such as hockey, ski jumping, and rugby. In North America, the professional Women's National Basketball Association (WNBA) and National Women's Soccer League (NWSL) are commercially viable businesses that emerged as a direct consequences of Title IX legislation.

One big reason Title IX worked was because it did not attempt to lump males and females into one pot. Imagine if it had, for example, by requiring that football and volleyball teams all have 50–50 membership by sex. While a mix like this works well for casual intramurals, in the most competitive leagues where winning is everything, the sex tendency in height and the sex difference in strength would work against women, not help them. A similar recognition that sex is "a thing" is needed for robust solutions in other, more fundamental areas as well. Let's look at a few:

- *Real and implied violence against women.* We cannot dismiss the idea that throughout history, men have dominated women largely because they can. The implied or actual use of force is a powerful advantage. Of course, it is used all too often, with lethal results. The news suggests modern American society is relying on the reactive responses of schools, employers, and law enforcement to promote a safe environment for women. This doesn't help victims—we need the be proactive. Restraining orders are one tool, but so is education. Central to any practical solution is establishing an environment in which women are respected, and males are trained in restraint from an early age. That is, we need to focus also on raising boys to become gentle men.
- *Objectification.* We evolved from seasonal breeding ancestors to become a species with year-round sexual activity and long-term pair bonding. The resulting mutual mate choice this imposes, combined with an earlier shift in sex discrimination from the olfactory to the visual, selected for bodies that are richly ornamented with cues about sex and fertility. The tendency to "check each other out" thus runs very deep and is somewhat involuntary. I doubt I'm the only person who has had to carefully file away spontaneous observations about a colleague's physique that are

irrelevant to our working relationship. I've never lived in a woman's head, but we males are constantly noting, evaluating, and sometimes fantasizing about the bodies of the people of the sex to whom we are generally attracted. This is not something to be celebrated or indulged, but it is involuntary. Left unchecked, however, it dehumanizes people (especially women). Once we understand this tension, we can focus on mitigating its worst consequences with education and smart policies. In my area of science, conference organizers now require all participants to agree to refrain from any sexual commentary or harassment of fellow participants, and it seems to be working. Other proposals are harder to judge. For example, are dress codes that promote "modesty" protective or paternalistic? This will require more discussion.

- *Pornography.* The visual focus of human sexual attraction makes us innately susceptible to porn. Before writing this book, I assumed that women rarely consume it, but data from the website Pornhub suggest about a quarter of its viewers are women.[4] Much like the failure of Prohibition, trying to ban porn will never work. For both industry participants and consumers, however, there are negative consequences. Actors are often exploited, and what gets filmed is not necessarily pleasurable in reality. If there is value in limiting porn consumption (and production), then perhaps we can start by acknowledging that we are being manipulated by our sensory biases, similar to the way potato chips are hard to resist. Such self-awareness is likely to be more effective in the long run than making people feel guilty.

- *Sexism.* There are plenty of bogus claims for sex differences that are not even sex tendencies. In acknowledging that male and female are meaningful categories of human beings, we need to be ever-mindful of the temptation to overgeneralize. When one senses this is happening, demanding to see real data with appropriate controls never hurts.

- *Sex education I: the basics.* A common theme of conservative takes on sex is that it should be strictly coupled to reproduction. This may apply to mice or lions, but the decoupling of sexual activity from reproduction is an ancient feature of the great apes (figure E.1). We humans have used technology and our knowledge of cause and effect to exaggerate this disconnection somewhat, but the social role of sexuality is as much a part of our history as upright walking. There are indeed risks associated

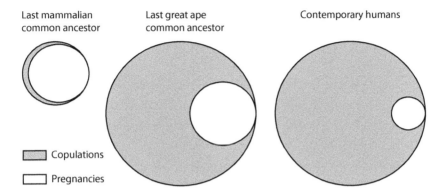

E.1 Comparison of different mammals regarding sexual activity, reproduction, and their relationships. Sexual interactions are more frequent in the great apes and humans (as indicated by their larger gray circles), but largely disconnected from reproduction. In the contemporary human societies this effect is exaggerated further.

with sexual activity: sexually transmitted diseases, unwanted pregnancy, public shaming or job loss, and intimate partner violence are all possible. So, should we just tell our kids that they should abstain until they are in a stable relationship, and not ask questions? This is nice and simple, but it will never work. After breathing and the urge to pee with a full bladder, our sex drive is one of humanity's most powerful urges. It appears at puberty unbidden and tends to get channeled somehow. Not talking about it doesn't prevent it from existing, and enforced ignorance and stigma groom the surest path to the worst consequences of sexual activity. In contrast, frank discussions based on accurate information demystify sex, and this approach actually delays the onset of sexual activity rather than accelerating it.[5]

- *Sex education II: sexual orientation.* In 2022, the news was filled with stories of states legislating against the mere acknowledgment that not everyone is straight or cisgendered in the context of school curricula.[6] The fear seems to be that by mentioning the variety of human sexual expressions, straight kids will be converted (infected?). The evidence reviewed in chapter 13 makes it clear that most queer adults knew they were different at an early age, an age well before puberty. Thinking back to my childhood, those kids (especially the boys) were sometimes identifiable and mercilessly harassed starting in elementary school. Much of this resulted from the failure to acknowledge the normalness of their presence. There is nothing frightening about teaching even young kids that we come in different flavors. If your kid is destined to be gay, what they learn in school won't stop it. If your kid is destined to be straight, what they learn in school won't change it. Most important is that this knowledge has the potential to reduce fear of the other; normalizing queerness in a way that keeps people safe.

- *Health care.* Most medical research was historically performed on men, and not always for reasons of chauvinism. In the 1960s, the drug thalidomide was often prescribed in Europe to combat morning sickness, but it was subsequently found to cause horrific defects in fetal development that truncated the arms and legs.[7] With these unfortunate babies in mind, in 1977, the U.S. Food and Drug Administration (FDA) recommended excluding women from early-phase clinical trials to prevent similar tragedies.[8] However, exclusion of females was also justified by the concern that physiological variation related to the menstrual cycle made data from women "noisy."[9] Not surprisingly, this introduced substantial gaps in medical knowledge, in which sex differences made male-specific data a poor predictor of female responses. In 2001, the Public Health Service Act reversed the pendulum and mandated that all National Institutes of Health (NIH) research include both male and female human subjects. As the NIH's own website notes, "The primary goal of this law is to ensure that research findings can be generalizable to the entire population."[10] The spirit of this act has also spread to research with nonhuman animals, such as mice. The importance of recognizing sex differences has led to better training for doctors and better diagnostic methods for common maladies, such as heart disease.[11]

- *Legal status of intersex people.* Sexual development occasionally produces an outcome that is not standard-issue male or female, often due to mutations that impact the genes that regulate it (see chapter 12). This is separate from the issues of how and when to acknowledge transgender identity, but the two are sometimes lumped together by activists and anti-activists alike. While gender identity will continue to spark controversy in the years ahead, we should all be able to agree today that there is no justification for forcing intersex people to choose between categories

that do not apply to them. To avoid that, all legal documents in which
biological sex is an obligatory component (like driver licenses, medical records, school enrollment forms, etc.) should offer "intersex" as an
option. Even if it only applies to a small fraction of the population, it is
the right thing to do.

- *The pay gap.* On average, men still earn more than women, and this is true
  even when comparing workers within a job type. One hypothesis might
  be that men are dominating women as a group because they can, perhaps
  due to unspoken physical intimidation. If this were true, we expect that
  the wage gap will apply similarly to women independent of the details
  of their personal lives. In 1980, this was largely the case. Today, however,
  80 percent of the salary difference between men and women can be
  attributed to the birth of a child.[12] This is still sexism, but it cannot be
  explained by a simple narrative of male power based on strength. Consistent with a shift from "being a female penalty" in the recent past to a
  "being a mother penalty" today, the net worth of childless U.S. women
  age fifty-five and older in 2018 was not only greater than that of mothers
  but also above that of childless men and fathers (though not significantly
  so in the latter case).[13] This may be a first glimpse of a gender-equitable
  world in which women's material rewards can match or exceed those of
  men in the right circumstances. However, it is telling that fathers do not
  experience a parent penalty. New mothers are more likely to cut their
  hours and to shift to "family friendly" (with lower pay) careers than
  are new fathers. This suggests that policies promoting greater paternal
  involvement and better access to early childhood day care will make all
  the difference. We can do this.

It would certainly have been simpler had we evolved to be hermaphroditic
instead, but that train had left the station by the Devonian. Thousands of articles
and hundreds of books have been published about sex and sexes, and countless
debates have been held about their significance for humanity. Thus, I do not
expect that this book will really settle anything. What I do sincerely hope, however, is that in understanding how we got here, both evolutionarily and developmentally, we will see more clearly how to work skillfully with our sexual selves.

Acknowledgments

After decades of writing focused on academic articles, I found writing a book for nonspecialists to be a different beast entirely. Its conception and completion were only possible because of the kindness, time, and openness to sometimes difficult topics of many people. I would like to acknowledge my gratitude to some of them here.

My late parents, Douglas and Betty Haag, encouraged me to be both curious and attentive to detail (the latter not coming naturally to me at all). As an undergraduate in the late 1980s, I was fortunate to study biology in a liberal arts setting, alongside the arts and humanities in a passionately idealistic community. This promoted looking for connections between historically siloed topics, and this book is an exercise in seeing where that leads us. I am especially grateful to two Oberlin College biologists, Yolanda Cruz and Catherine McCormick, for their encouragement both before and after my graduation.

I was similarly fortunate to spend my graduate career in the Department of Biology at Indiana University, Bloomington, a bastion of integrative biology. My late doctoral adviser, Rudy Raff, was an inspiration in both the conduct of interdisciplinary evolutionary biology and as an author of important books. From my postdoctoral mentor at the University of Wisconsin, Judith Kimble, I gained additional skills that helped me become truly independent. Since coming to the University of Maryland in 2002, my colleagues and students have provided a wonderfully fertile environment in which I could further develop as a scholar, teacher, and researcher. I'd like to acknowledge Eric Baehrecke, Alexandra Bely, Karen Carleton, William Jeffery, Antony Jose, Scott Juntti, Thomas

Kocher, Carlos Machado, Stephen Mount, Leslie Pick, and Louisa Wu for their especially big roles.

When this book was still an early embryo whose eventual form was still unclear, I received excellent advice from some seasoned experts, including Sean B. Carroll and Russell Galen. I am grateful for their advice on the scope of the book and overall style. The anonymous reviewers of the initial book proposal also played an important role in shaping the content. Further commitment to the project came from a colorectal cancer diagnosis in 2014. Nothing motivates like a stark reminder of one's mortality, and I will always be grateful to the skilled doctors who quite literally saved my ass.

As I delved into topics that were new to me, I benefitted greatly from discussions with experts, including Daniel Blackburn, Dave Grossnickle, Richard Michod, Mihaela Pavlicev, Paul Vasey, and Lucas Weaver. Many others contributed through briefer conversations, and I am sorry not to be able to name everyone here. Once ideas coalesced into chapter drafts, early feedback on some of them came from Alexa Bely, Matt Giorgianni, Elisabeth Newcomb, the Nerd Night crew, Nitish Sinha, Justin Van Goor, and David Zarkower.

Telling a story of this breadth requires merging newly created text and graphics with those from a range of other sources. In producing original art to communicate key ideas, I relied greatly on Nick Bezio as a gifted and always patient partner. In granting access to unpublished textual and biological resources, I am especially grateful to Erin Harbour at the Dolph Briscoe Center for American History, University of Texas at Austin, and to Elizabeth Lockett of the Human Developmental Anatomy Center. In addition, a number of biology colleagues and friends freely shared unpublished images, including Jean-Bernard Caron, Charles Delwiche, Richard Elinson, Steffen Harzsch, Jih-Pai Lin, Tim Maugel, Nico Michiels, Sam Patania, Tom Shannon III, and Michael Vanden Berg.

To tweak a legal adage, the writer who relies on themself as their editor has a fool for a client. Given the book's often-sensitive subject matter, this is especially true here. Lisa Diamond, Rebecca Haag, and two anonymous reviewers provided a host of helpful reactions to the complete draft manuscript, which led to many changes that greatly strengthened the text. I am also heavily indebted to Sara Carminati for pointing me toward important sources that I had overlooked and in helping me better appreciate the different ways in which our choices of words can be either welcoming or exclusionary. Ethan Mereish provided additional suggestions for inclusive language.

I am extraordinarily grateful to Miranda Martin at Columbia University Press for her willingness to take on the project and for her sensitive guidance along the way. This allowed a dream (born of something like a midlife crisis) to become a reality.

Finally, over the years required to complete this project, my spouse, children, and research trainees had to live through the process with me. I am thankful that we all seem to have survived it.

Glossary

ALTERNATIVE REPRODUCTIVE TACTIC (ART) Reproductive behaviors that differ from that which is more commonly observed in a species, most often in males that avoid direct male-male competition. ARTs are usually triggered by environmental conditions.

AMNIOTE Any vertebrate animal that possesses an egg with a specialized membrane (the amnion) that surrounds the embryo, providing protection and allowing for development on land; includes lizards, snakes, turtles, birds, and mammals.

AMPHIBIAN A tetrapod vertebrate with external fertilization and lacking membranes or shell around its eggs; amphibians are dependent on water for reproduction.

ANDROPHILIA Sexual orientation toward males/men.

ANISOGAMY Existence of a large (egg) and small (sperm) gamete type in a species.

ARCHAEA One of the two domains of prokaryotic life; mostly heterotrophic and oxygen-intolerant (anaerobic), and presently live in extreme environments.

ASEXUAL In biology: reproduction through means other than fertilization of two haploid gametes, including parthenogenic embryos, budding, etc. In humans: an identity characterized by no inclination toward sexual activity.

AUTOSOME Any chromosome whose presence and/or number is identical between sexes.

AUTOTROPH Greek for "self-feeder;" an organism that produces its own food from atmospheric carbon dioxide and an energy source, most commonly by the process of photosynthesis, whose energy source is sunlight.

BACTERIA One of the two domains of prokaryotic life. They include both autotrophic and heterotrophic forms, and most are oxygen-tolerating (aerobic).

BARR BODY The largely inactivated copy of the X chromosome in female mammals.

CISGENDER A human identity in which natal biological sex (male or female) is aligned with gender, that is, a female who identifies as a girl or woman, or a male who identifies as a boy or man.

CLITORIS The eventual derivative of the embryonic genital tubercle in female mammals.

CROSSING OVER The physical exchange of DNA between chromosomes that occurs during the first meiotic division; creates a chromosome with novel combinations of genetic variants that were not found in either of the two progenitors. *Also known as* recombination.

CYANOBACTERIA Phylum of photosynthetic bacteria that were among the first living organisms on Earth; the "cyan" part of their name refers to their color, and they have been commonly called blue-green algae, despite being prokaryotic.

DECIDUA Vascularized tissue within the mammalian uterus that forms a site of implantation for an embryo.

DIFFERENCE OF SEXUAL DEVELOPMENT In humans, a congenital state in which one or more aspects of sexual anatomy are atypical; a broader category than intersex.

DIPLOID Possessing two copies of the genome, including all autosomes.

DOUBLESEX-MAB-3 (DM) PROTEINS Family of DNA-binding transcription factor proteins associated with male development across all animals; named for founding examples from *Drosophila* and *C. elegans*, respectively.

ELECTRON TRANSPORT CHAIN (ETC) A complex of proteins imbedded in the membranes of prokaryotes and of the eukaryotic organelles that are descended from them (mitochondria and chloroplasts); ETCs, powered by a source of chemical energy, produce energy-carrying molecules (e.g., ATP) that fuel cell work.

ESTROGEN A steroid sex hormone associated with female development and physiology in vertebrates but is also produced by males.

EUKARYA Domain of life formed by the union of an archean and an aerobic bacterium, the latter of which became the mitochondrion; only eukaryotes have true sex, that is, genome-scale meiotic recombination alternating with syngamy.

EUTHERIA Placental mammals, that is, those with extended pregnancy and no pouch.

EXOGAMY A cultural practice or rule that mating partners must be chosen from outside a particular social group, tribe, or clan; helps to avoid close kinship marriages and encourages social connections and alliances with other groups.

FA'AFAFINE Androphillic Samoans who are born male but adopt female dress and social roles.

GAMETE In eukaryotes, a haploid cell that has the capacity to fuse (under fertilization) with another such cell.

GAY Same-sex or same-gender sexual orientation.

GENDER The roles, behaviors, and identities attributed to individuals that relate to their perceived sex.

GENDER DYSPHORIA A psychological state of distress in which a person's internal sense of their gender does not match their biological sex at birth.

GENITAL TUBERCLE The embryonic structure of mammals that gives rise to either the glans of the penis (in males) or the clitoris (in females).

GONADOTROPIN Small protein hormones produced by the anterior pituitary of all vertebrates, including follicle-stimulating hormone (FSH) and luteinizing hormone (LH); a version of LH is also produced by the placenta in primates and horses.

GUBERNACULUM Cord-like structure that serves as a guide and support for the movement of the testes from their original position in the abdomen to their proper location within the scrotum.

GYNOPHILIC Sexual orientation toward females/women.

HAPLOID Possessing a single copy of the genome.

HERMAPHRODITE A single organism that produces both female and male gametes (inappropriately used in the past to refer to intersex people).

HETEROTROPH Greek for "feeds on others;" an organism that acquires its food by consuming the living or dead cells of other organisms; for herbivorous heterotrophs, the other organisms being consumed are autotrophs.

HOMOZYGOTE A diploid organism for which both copies of a particular genomic site are identical.

INBREEDING DEPRESSION Reduced health or vigor due to mating between close relatives, or to self-fertilization; caused by homozygosity of recessive mutations.

INDEPENDENT ASSORTMENT A feature of meiosis in which an individual chromosome from the diploid set is distributed to a haploid product regardless of the origin of the other chromosomes in that product.

INTERSEX A condition in which anatomical, genetic, and/or hormonal features that typify each binary sex are instead intermediate between male and female states. Intersex is one type of difference of sexual development.

ISOGAMY Formation of gametes of equal size.

KIN SELECTION A form of natural selection in which seemingly altruistic traits that reduce or eliminate reproduction are indirectly adaptive because of a benefit they provide to the reproductive success of close relatives.

LABIA MAJORA The larger, outer vaginal lips of female mammals; developmentally equivalent to the scrotum of males.

LEYDIG CELLS A cell type of mammalian testes, found between the sperm-producing seminiferous tubules; respond to luteinizing hormone by producing testosterone.

LOKIARCHAEOTA Phylum of Archaea most closely related to the eukaryote lineage; possess proteins and cell features previously thought to be specific to eukaryotes.

MEIOSIS Specialized form of cell division unique to eukaryotes, in which a diploid progenitor cell produces four haploid descendants, each with a unique combination of genetic variants.

MENARCHE In women, the age at which the first ovulation and menstrual period occur.

MENOPAUSE In women, the phase of life after cessation of ovulation.

METATHERIA Mammals with a brief pregnancy and a pouch, that is, marsupials.

MITOCHONDRION Endosymbiotic organelle within eukaryotic cells, the site of most energy capture from food (using a process called oxidative phosphorylation) and of oxygen consumption; once free-living aerobic bacteria.

MONOGAMY Animal mating system in which males and females form exclusive (or nearly exclusive) pair bonds for at least one breeding season.

MÜLLERIAN DUCT Embryonic tube that persists in female mammals to form the oviducts and uterus but is lost in males.

NONBINARY A gender identity that is neither consistently masculine nor consistently feminine.

OVULATION The release of a mature egg from the ovary.

PARTHENOGENESIS Greek for "virgin origin;" a form of asexual reproduction by females that occurs without a mate; generally produces offspring that are genetically identical to the mother.

PENIS The organ of male amniotes that allows internal deposition of sperm into the female reproductive tract.

PHYLOGENY A branching diagram depicting the inferred evolutionary relationships of a set of organisms.

PLACENTA In live-bearing mammals, the organ that forms an interface with the uterus for the exchange of gases, nutrients, and wastes between mother and developing offspring; derived from embryonic cells.

POLYANDRY An animal mating system in which one female has multiple males as mates.

POLYGYNY An animal mating system in which one male has multiple females as mates, often to the exclusion of other males.

POLYPHENISM Discrete alternative states produced during development under the influence of environmental inputs and not under genetic control.

PSEUDO-AUTOSOMAL REGION (PAR) The portion of a sex chromosome that recombines freely with the alternative sex chromosome, as autosomes do.

QUEER Holding a sexual orientation or gender identity other than strictly heterosexual and/or cisgender; generally broader than "gay."

REACTIVE OXYGEN SPECIES (ROS) Chemical by-products of mitochondrial function that damage cell components.

RECOMBINATION *See* crossing over.

SCROTUM The pouch into which the testes of most mammals are pulled during development; the counterpart of the labia majora in females.

SELF-FERTILITY Form of one-parent reproduction found in some hermaphrodites in which an egg is fertilized by the sperm of the same organism.

SEQUENTIAL HERMAPHRODITE An individual that changes from one biological sex to another over the course of its life.

SERTOLI CELLS Cells of the mammalian testis that provide structural and nutritional support to developing sperm cells as they go through various stages of maturation.

SEX (1) A characteristic of an organism defined by the type of gamete produced, with the sex of sperm producers being male, the sex of egg producers being female, and the sex of those producing both being hermaphrodic; (2) the entire process of sexual reproduction, including meiosis, the formation of haploid gametes, and fertilization to produce diploids; (3) the act of mating or genital stimulation in animals.

SEX CHROMOSOME A chromosome that varies in form or number between the sexes and often directly controls sex determination.

SEX DETERMINATION The process by which an initially indifferent organism is fated to become male, female, or hermaphrodite; can be regulated by genes or environment.

SEX DIFFERENCE A trait that categorically distinguishes the sexes of a species, that is, an essentially binary feature.

SEX TENDENCY A trait that does not categorically distinguish sexes but which has a significantly different average or most-common value (mode) between sexes.

SEXUAL SELECTION Differential reproductive success based on differential access to mates.

SOX9 Vertebrate gene that encodes a transcription factor of the high-mobility group (HMG) family, related to *SRY*; in mammals, it is an autosomal gene that promotes male development.

SRY A mammalian gene encoding a transcription factor related to SOX9 that is found on the Y chromosome and constitutes the testis-determining factor; promotes male development.

SYNGAMY General term for gamete fusion; synonymous with "fertilization" in anisogamous organisms.

TESTOSTERONE Steroid sex hormone associated with male development and fertility in vertebrates but is also produced in females.

TRANSGENDER A range of gender identities that do not match those typically associated with one's biological sex.

TRANSCRIPTION FACTOR A protein that regulates the activity of genes, usually by directly binding the target's DNA.

TRANSSEXUAL An outmoded term used historically to refer to a person who has altered their anatomy, through hormones and/or surgery, to resemble that of the opposite biological sex; considered offensive to most transpeople.

URETHRA In mammals, the tube that connects the urinary bladder to the exterior of the body; in males, the portion within the penis also delivers sperm.

VAS DEFERENS In male mammals, the tube that conveys mature sperm from each testicle to the urethra. The vas deferens is severed in a vasectomy.

VOMERONASAL ORGAN (VNO) In amniotes, a sensory organ that is specialized for detection of pheromones.

WOLFFIAN DUCT In mammals, an embryonic tube that persists in males to form the epididymis and vas deferens but is lost in females.

ZYGOTE The diploid cell that results from syngamy/fertilization.

Notes

INTRODUCTION

1. Mansplaining is the tendency many men have to lecture to women in a condescending or patronizing manner, often assuming that they have less knowledge or understanding about the topic (from "man" + "explaining").
2. Because a scientist's training is very personal, I owe a great deal to the people that gave me a chance to develop this approach through working alongside them. These dedicated people are named in the acknowledgments.

1. SEX IN THE AIR

1. An excellent tool for visualizing changes in our atmosphere over time is Earth Viewer, from the Howard Hughes Medical Institute. See http://media.hhmi.org/biointeractive /earthviewer_web/earthviewer.html.
2. M. R. Walter, R. Buick, and J. S. R. Dunlop, "Stromatolites 3,400–3,500 Myr Old from the North Pole Area, Western Australia," *Nature* 284, no. 3 (April 1980): 443–45, https:// doi.org/10.1038/284443a0.
3. Keeping these terms straight is easier if their Greek origins are kept in mind. The word *karyon* means "kernel" and refers to the nucleus. *Eu*karyotes have a "*true* kernel," while the *pro*karyotes are named for being "*before* the kernel." Although named from examination of present-day organisms, it turns out that prokaryotes really did come first in evolution, as will be clear by the end of this chapter.
4. E. Kallee and S. Okada, "Effect of X-Rays on Mitochondrial Desoxyribonuclease II," *Experimental Cell Research* 11, no. 1 (August 1956): 212–214, https://doi.org/10.1016/0014 -4827(56)90206-3; S. Okada and L. D. Peachy, "Effect of Gamma Irradiation on the Desoxyribonuclease II Activity of Isolated Mitochondria," *Journal of Biophysical and Biochemical Cytology* 3, no. 2 (March 1957): 239–48, 239–48. https://doi.org/10.1083/jcb.3.2.239.
5. M. M. Nass and S. Nass, "Intramitochondrial Fibers with DNA Characteristics. I. Fixation and Electron Staining Reactions," *Journal of Cell Biology* 19 (December 1963): 593–611, https://doi.org/10.1083/jcb.19.3.613.

6. E. Guttes and S. Guttes, "Thymidine Incorporation by Mitochondria in Physarum Polycephalum," *Science* 145, no. 3636 (September 1964): 1057–8, https://doi.org/10.1126/science.145.3636.1057.

7. L. Sagan, "On the Origin of Mitosing Cells," *Journal of Theoretical Biology* 14, no. 3 (March 1967): 255–274, https://doi.org/10.1016/0022-5193(67)90079-3. Some prominent biologists at the time initially found endosymbiosis too far-fetched. Among them was my own beloved doctoral mentor, Rudolf Raff, but eventually he was won over by mounting evidence.

8. Plant mitochondrial genomes, while encoding a similar number of genes as their animal counterparts, are over ten times bigger and not always circular. The reasons for this difference are not clear.

9. C. R. Woese and G. E. Fox, "Phylogenetic Structure of the Prokaryotic Domain: The Primary Kingdoms," *Proceedings of the National Academy of Sciences of the United States of America* 74, no. 11 (November 1977): 5088–5090, https://doi.org/10.1073/pnas.74.11.5088.

10. N. Iwabe, K. Kuma, M. Hasegawa, S. Osawa, and T. Miyata, "Evolutionary Relationship of Archaebacteria, Eubacteria, and Eukaryotes Inferred from Phylogenetic Trees of Duplicated Genes," *Proceedings of the National Academy of Sciences of the United States of America* 86, no. 23 (December 1989): 9355–9359, https://doi.org/10.1073/pnas.86.23.9355.

11. J. A. Lake, E. Henderson, M. Oakes, and M. W. Clark, "Eocytes: A New Ribosome Structure Indicates a Kingdom with a Close Relationship to Eukaryotes," *Proceedings of the National Academy of Sciences of the United States of America* 81, no. 12 (June 1984): 3786–3790, https://doi.org/10.1073/pnas.81.12.3786.

12. A. Spang, J. H. Saw, S. L. Jorgensen, K. Zaremba-Niedzwiedzka, J. Martijn, A. E. Lind, R. van Eijk, et al, "Complex Archaea That Bridge the Gap Between Prokaryotes and Eukaryotes," *Nature* 521, no. 7551 (May 2015): 173–179, https://doi.org/10.1038/nature14447. Loki is a shape-shifting character in ancient Norse mythology. His name was given to a nearby hydrothermal vent, Loki's Castle.

13. K. Zaremba-Niedzwiedzka, E. F. Caceres, J. H. Saw, D. Backstrom, L. Juzokaite, E. Vancaester, K. W. Seitz, et al., "Asgard Archaea Illuminate the Origin of Eukaryotic Cellular Complexity," *Nature* 541, no. 7637 (January 19 2017): 353–358, https://doi.org/10.1038/nature21031.

14. H. Imachi, M. K. Nobu, N. Nakahara, Y. Morono, M. Ogawara, Y. Takaki, Y. Takano, et al., "Isolation of an Archaeon at the Prokaryote-Eukaryote Interface," *Nature* 577, no. 7791 (January 2020): 519–525, https://doi.org/10.1038/s41586-019-1916-6.

15. Prometheus, you may recall, was the Greek god that formed humans from mud, making an apt namesake for a close relative of the protoeukaryote.

16. W. Martin and M. Muller, "The Hydrogen Hypothesis for the First Eukaryote," *Nature* 392, no. 6671 (March 1998): 37–41, https://doi.org/10.1038/32096.

17. D. Speijer, "Alternating Terminal Electron-Acceptors at the Basis of Symbiogenesis: How Oxygen Ignited Eukaryotic Evolution," *BioEssays* 39, no. 2 (February 2017), https://doi.org/10.1002/bies.201600174.

18. E. Hagstrom and S. G. Andersson, "The Challenges of Integrating Two Genomes in One Cell," *Current Opinion in Microbiology* 41 (February 2018): 89–94, https://doi.org/10.1016/j .mib.2017.12.003.

19. R. E. Michod and B. R. Levin, eds., *The Evolution of Sex: An Examination of Current Ideas* (Sunderland, MA: Sinauer, 1988).

20. K. von Besser, A. C. Frank, M. A. Johnson, and D. Preuss, "*Arabidopsis HAP2 (GCS1)* Is a Sperm-Specific Gene Required for Pollen Tube Guidance and Fertilization," *Development* 133, no. 23 (December 2006): 4761–4769, https://doi.org/10.1242/dev.02683; T. Mori, H. Kuroiwa, T. Higashiyama, and T. Kuroiwa, "Generative Cell-Specific 1 Is Essential for Angiosperm Fertilization," *Nature Cell Biology* 8, no. 1 (January 2006): 64–71, https://doi .org/10.1038/ncb1345.

21. G. Bloomfield, "The Molecular Foundations of Zygosis," *Cellular and Molecular Life Sciences* 77, no. 2 (January 2020): 323–330, https://doi.org/10.1007/s00018-019-03187-1.

22. J. R. True and E. S. Haag, "Developmental System Drift and Flexibility in Evolutionary Trajectories," *Evolution & Development* 3, no. 2 (March–April 2001): 109–119, https://doi .org/10.1046/j.1525-142x.2001.003002109.x

23. J. Fedry, Y. Liu, G. Pehau-Arnaudet, J. Pei, W. Li, M. A. Tortorici, F. Traincard, et al., "The Ancient Gamete Fusogen Hap2 Is a Eukaryotic Class Ii Fusion Protein," *Cell* 168, no. 5 (February 2017): 904–1015e10, https://doi.org/10.1016/j.cell.2017.01.024.

24. See, for example, W. G. Hill and A. Robertson, "The Effect of Linkage on Limits to Artificial Selection," *Genetics Research* 8, no. 3 (December 1966): 269–294; A. J. Betancourt and D. C. Presgraves, "Linkage Limits the Power of Natural Selection in *Drosophila*," *Proceedings of the National Academy of Sciences of the United States of America* 99, no. 21 (October 2002): 13616–13620, https://doi.org/10.1073/pnas.212277199.

25. R. E. Michod and B. R. Levin, *The Evolution of Sex* (Sunderland, MA: Sinauer, 1988).

26. J. D. Macdonald, "On the Structure of the Diatomaceous Frustule, and Its Genetic Cycle," *Annals and Magazine of Natural History*, 3, no. 13 (January 1869): 1–8, https://doi .org/10.1080/00222936908695866.

27. E. Pfitzer, "Untersuchungen über Bau und Entwickelung der Bacillariaceen (Diatomaceen)," *Botanische Abhandlungen aus dem Gebiet der Morphologie und Physiologie* 1, no. 2 (1871): 1–189.

28. L. Geitler, *Der Formwechsel der pennaten Diatomeen* (Jena: Gustav Fischer, 1932).

29. U. Goodenough, H. Lin, and J. H. Lee, "Sex Determination in Chlamydomonas," *Seminars in Cell and Developmental Biology* 18, no. 3 (June 2007): 350–361, https://doi.org/10.1016/j .semcdb.2007.02.006.

30. E. S. Haag, "Why Two Sexes? Sex Determination in Multicellular Organisms and Protistan Mating Types," *Seminars in Cell and Developmental Biology* 18, no. 3 (June 2007): 348–349, https://doi.org/10.1016/j.semcdb.2007.05.009

31. F. Ishikawa and T. Naito, "Why Do We Have Linear Chromosomes? A Matter of Adam and Eve," *Mutation Research* 434, no. 2 (June 1999): 99–107, https://doi.org/10.1016 /s0921-8777(99)00017-8.

32. J. L. Wong and M. A. Johnson, "Is Hap2-Gcs1 an Ancestral Gamete Fusogen?," *Trends in Cell Biology* 20, no. 3 (March 2010): 134–141, https://doi.org/10.1016/j.tcb.2009.12.007.

2. EX UNUM PLURIBUS

1.   In developmental biology, an embryo is generally any stage from fertilization until hatching. In humans, however, there is a tendency to use the term "embryo" for the first trimester only. At the end of this stage, the body's organs have formed in their rudimentary state. "Fetus" is used for all later prebirth development, which is mainly characterized by the growth of structures formed in the embryo. As we will see, however, for sexual organs, several major events occur in the fetal phase.

2.   J. Maienschein, M. Glitz, and G. E. Allen, *Centennial History of the Carnegie Institution of Washington: Volume V, The Department of Embryology* (Cambridge: Cambridge University Press; 2004).

3.   Many of the Carnegie sections are available online via The Virtual Human Embryo resource, https://www.ehd.org/virtual-human-embryo/.

4.   As will be discussed in chapter 12, there are exceptions to the XX female, XY male pattern, which lead to intersex conditions that don't fit neatly into this paradigm.

5.   E. Rajpert-De Meyts, N. Jorgensen, N. Graem, J. Muller, R. L. Cate, and N. E. Skakkebaek, "Expression of Anti-Mullerian Hormone During Normal and Pathological Gonadal Development: Association with Differentiation of Sertoli and Granulosa Cells," *Journal of Clinical Endocrinology and Metabolism* 84, no. 10 (October 1999): 3836–3844, https://doi.org/10.1210/jcem.84.10.6047.

6.   S. E. Hannema and I. A. Hughes, "Regulation of Wolffian Duct Development," *Hormone Research* 67, no. 3 (2007): 142–151, https://doi.org/10.1159/000096644. https://www.ncbi.nlm.nih.gov/pubmed/17077643.

7.   N. Berkane, P. Liere, J. P. Oudinet, A. Hertig, G. Lefevre, N. Pluchino, M. Schumacher, and N. Chabbert-Buffet, "From Pregnancy to Preeclampsia: A Key Role for Estrogens," *Endocrine Reviews* 38, no. 2 (April 2017): 123–144, https://doi.org/10.1210/er.2016-1065.

8.   L. Baskin, J. Shen, A. Sinclair, M. Cao, X. Liu, G. Liu, D. Isaacson, et al., "Development of the Human Penis and Clitoris," *Differentiation* 103 (September–October 2018): 74–85, https://doi.org/10.1016/j.diff.2018.08.001.

9.   J. Shen, G. R. Cunha, A. Sinclair, M. Cao, D. Isaacson, and L. Baskin, "Macroscopic Whole-Mounts of the Developing Human Fetal Urogenital-Genital Tract: Indifferent Stage to Male and Female Differentiation," *Differentiation* 103 (September–October 2018): 5–13, https://doi.org/10.1016/j.diff.2018.08.003.

10.   G. R. Cunha, S. J. Robboy, T. Kurita, D. Isaacson, J. Shen, M. Cao, and L. S. Baskin, "Development of the Human Female Reproductive Tract," *Differentiation* 103 (September–October 2018): 46–65, https://doi.org/10.1016/j.diff.2018.09.001.

11.   V. Sharma, T. Lehmann, H. Stuckas, L. Funke, and M. Hiller, "Loss of Rxfp2 and Insl3 Genes in Afrotheria Shows That Testicular Descent Is the Ancestral Condition in Placental Mammals," *PLoS Biology* 16, no. 6 (June 2018): e2005293, https://doi.org/10.1371/journal.pbio.2005293; L. Werdelin and A. Nilsonne, "The Evolution of the Scrotum and Testicular Descent in Mammals: A Phylogenetic View," *Journal of Theoretical Biology* 196, no. 1 (January 1999): 61–72, https://doi.org/10.1006/jtbi.1998.0821.

12.   S. Shefi, P. E. Tarapore, T. J. Walsh, M. Croughan, and P. J. Turek, "Wet Heat Exposure: A Potentially Reversible Cause of Low Semen Quality in Infertile Men," *International Brazilian Journal of Urology* 33, no. 1 (January–February 2007): 50–56, discussion 56–7, https://doi.org/10.1590/s1677-55382007000100008.

13. G. Warnke, *Debating Sex and Gender* (Oxford: Oxford University Press, 2011).

14. Women do produce some testosterone in their ovaries and adrenal glands, and this has a role in sexual desire and mood. Men produce estrogen as a necessary intermediate in the synthesis of testosterone, and their fertility depends directly on the estrogen for maintaining fluid balance in the testes. See A. Guay and S. R. Davis, "Testosterone Insufficiency in Women: Fact or Fiction?," *World Journal of Urology* 20, no. 2 (June 2002): 106–110, https://doi.org/10.1007/s00345-002-0267-2; and Q. Zhou, L. Clarke, R. Nie, K. Carnes, L. W. Lai, Y. H. Lien, A. Verkman, et al., "Estrogen Action and Male Fertility: Roles of the Sodium/Hydrogen Exchanger-3 and Fluid Reabsorption in Reproductive Tract Function," *Proceedings of the National Academy of Sciences of the United States of America* 98, no. 24 (November 2001): 14132–14137, https://doi.org/10.1073/pnas.241245898.

15. For example, see Center for Transgender and Gender Expansive Health, "Gender-Affirming Care," Johns Hopkins Medicine, https://www.hopkinsmedicine.org/center-transgender -health/.

16. K. Takatsu, K. Miyaoku, S. Roy, et al., "Induction of Female-to-Male Sex Change in Adult Zebrafish by Aromatase Inhibitor Treatment," *Scientific Reports* 3 (2013): 3400, https://doi.org/10.1038/srep03400.

3. GO BIG

1. S. A. F. Darroch, E. F. Smith, M. Laflamme, and D. H. Erwin, "Ediacaran Extinction and Cambrian Explosion," *Trends in Ecology & Evolution* 33, no. 9 (September 2018): 653–663, https://doi.org/10.1016/j.tree.2018.06.003.

2. J. G. Umen, "Green Algae and the Origins of Multicellularity in the Plant Kingdom," *Cold Spring Harbor Perspectives in Biology* 6, no. 11 (October 2014): a016170, https://doi.org/10.1101/cshperspect.a016170.

3. C. F. Delwiche and R. E. Timme, "Plants," *Current Biology* 21, no. 11 (June 2011): R417–R422, https://doi.org/10.1016/j.cub.2011.04.021; C. H. Wellman, "The Nature and Evolutionary Relationships of the Earliest Land Plants," *New Phytologist* 202, no. 1 (April 2014): 1–3, https://doi.org/10.1111/nph.12670; D. Edwards and P. Kenrick, "The Early Evolution of Land Plants, from Fossils to Genomics: A Commentary on Lang (1937) 'on the Plant-Remains from the Downtonian of England and Wales,'" *Philosophical Transactions of the Royal Society (London) B: Biological Sciences* 370, no. 1666 (April 2015), https://doi.org/10.1098/rstb.2014.0343.

4. X. Yuan, S. Xiao, and T. N. Taylor, "Lichen-Like Symbiosis 600 Million Years Ago," *Science* 308, no. 5724 (May 13 2005): 1017–20, https://doi.org/10.1126/science.1111347.

5. N. Eldridge, *Why We Do It* (New York: Norton, 2004).

6. W. B. Wood, ed., The Nematode *Caenorhabditis elegans* (Woodbury, NY: Cold Spring Harbor Laboratory Press, 1988); A. Hayward and J. F. Gillooly, "The Cost of Sex: Quantifying Energetic Investment in Gamete Production by Males and Females," *PLoS One* 6, no. 1 (January 2011): e16557, https://doi.org/10.1371/journal.pone.0016557.

7. E. S. Haag and T. W. Lo, "How to Make a Billion Parasites," *Developmental Cell* 45, no. 2 (April 2018): 147–148, https://doi.org/10.1016/j.devcel.2018.04.006.

8. A. Weismann, *Das Keimplasma: Eine Theorie der Vererbung*, trans. W. Newton Parker (1892); Harriet Rönnfeldt, *The Germ-Plasm: A Theory of Heredity* (New York: Scribner, 1893).

9.   G. A. Parker, R. R. Baker, and V. G. Smith, "The Origin and Evolution of Gamete Dimorphism and the Male-Female Phenomenon," *Journal of Theoretical Biology* 36, no. 3 (September 1972): 529–553, https://doi.org/10.1016/0022-5193(72)90007-0; G. Bell, "The Evolution of Anisogamy," *Journal of Theoretical Biology* 73, no. 2 (July 1978): 247–270. https://doi.org/10.1016/0022-5193(78)90189-3.

4. THE LEFT FIN OF DARKNESS

1.   A. Sekizawa, S. Satoko, T. Masakazu, S. Sakiko, and N. Yasuhiro, "Disposable penis and its replenishment in a simultaneous hermaphrodite," *Biology Letters* 9, no. 2 (April 2013), http://doi.org/10.1098/rsbl.2012.1150.

2.   N. Michiels and L. Newman, "Sex and Violence in Hermaphrodites," *Nature* 391, no. 647 (February 1998), https://doi.org/10.1038/35527.

3.   J. H. Callomon, "Sexual Dimorphism in Jurassic Ammonites," *Transactions of the Leicester Literary and Philosophical Society* 57 (1963): 20–56; H. Parent and M. Zatón, "Sexual Dimorphism in the Bathonian Morphoceratid Ammonite," *Polysphinctites tenuiplicatus*, *Acta Palaeontologica Polonica* 61, no. 4 (October 2016): 875–884.

4.   H. D. Pianka, "Ctenophora," in *Reproduction of Marine Invertebrates: Acoelomate And Pseudocoelomate Metazoans*, ed. A. C. Giese and J. S. Pearse (New York: Academic Press, 1974), 201–65; S. Sadro, "Porifera: The Sponges," in *An Identification Guide to Larval Marine Invertebrates of the Pacific Northwest*, ed. A. L. Shanks (Corvallis: Oregon State University Press, 2001), 5–12.

5.   A. M. Kerr, A. H. Baird, and T. P. Hughes, "Correlated Evolution of Sex and Reproductive Mode in Corals (Anthozoa: Scleractinia)," *Proceedings of the Royal Society B* 278, no. 1702 (January 2011): 75–81, https://doi.org/10.1098/rspb.2010.1196; A. C. Marques and A. G. Collins, "Cladistic Analysis of Medusozoa and Cnidarian Evolution," *Invertebrate Biology* 123 (May 2005): 23–42, https://doi.org/10.1111/j.1744-7410.2004.tb00139.x

6.   A. Weigert and C. Bleidorn, "Current Status of Annelid Phylogeny," *Organisms Diversity & Evolution* 16 (June 2016): 345–362, https://doi.org/10.1007/s13127-016-0265-7; T. J. Pandian, *Reproduction and Development in Annelida* (Boca Raton, FL: CRC Press, 2019).

7.   The smart person I'll lean on heavily here is Eric L. Charnov, specifically his book *The Theory of Sex Allocation* (Princeton, NJ: Princeton University Press, 1982).

8.   I have spent most of my career studying cases of selfing that evolved recently in otherwise male-female animals. It is an interesting puzzle, both in terms of how it comes into being and the consequences of adopting it. You can learn more in these published works: C. G. Thomas, G. C. Woodruff, and E. S. Haag, "Causes and Consequences of the Evolution of Reproductive Mode in Caenorhabditis Nematodes," *Trends in Genetics* 28, no. 5 (May 2012): 213–220, https://doi.org/10.1016/j.tig.2012.02.007; D. Yin, E. M. Schwarz, C. G. Thomas, R. L. Felde, I. F. Korf, A. D. Cutter, C. M. Schartner, et al., "Rapid Genome Shrinkage in a Self-Fertile Nematode Reveals Sperm Competition Proteins," *Science* 359, no. 6371 (January 5 2018): 55–61, https://doi.org/10.1126/science.aao0827.

9.   S. A. Ramm, A. Schlatter, M. Poirier, and L. Scharer, "Hypodermic Self-Insemination as a Reproductive Assurance Strategy," *Proceedings of the Royal Society B* 282 (July 2015): 20150660, https://doi.org/10.1098/rspb.2015.0660.

10.   W. Fu, R. M. Gittelman, M. J. Bamshad, and J. M. Akey, "Characteristics of Neutral and Deleterious Protein-Coding Variation among Individuals and Populations," *American Journal of Human Genetics* 95, no. 4 (October 2014): 421–436, https://doi.org/10.1016/j.ajhg.2014.09.006.

11.   E. S. Dolgin, B. Charlesworth, S. E. Baird, and A. D. Cutter, "Inbreeding and Outbreeding Depression in Caenorhabditis Nematodes," *Evolution* 61, no. 6 (June 2007): 1339–52, https://doi.org/10.1111/j.1558-5646.2007.00118.x.

12.   R. W. Harrington Jr., "Oviparous Hermaphroditic Fish with Internal Self-Fertilization." [in English]. *Science* 134, no. 3492 (December 1961): 1749–1750, https://doi.org/10.1126/science.134.3492.1749; M. Qu, S. Ding, M. Schartl, and M. C. Adolfi, "Spatial and Temporal Expression Pattern of Sex-Related Genes in Ovo-Testis of the Self-Fertilizing Mangrove Killifish (*Kryptolebias marmoratus*)," *Gene* 742 (June 2020): 144581, https://doi.org/10.1016/j.gene.2020.144581.

13.   R. Stott, *Darwin and the Barnacle* (New York: Norton, 2003).

14.   See, for example, C. Darwin, *A Monograph on the Sub-Class Cirripedia (The Balanidae and the Verrucidae)* (London: Ray Society, 1854).

15.   C. Darwin, *On the Origin of Species by Means of Natural Selection* (London: John Murray, 1859).

16.   E. R. Furman and A. B. Yule, "Self-Fertilisation in *Balanus improvisus* Darwin," *Journal of Experimental Marine Biology and Ecology* 144, nos. 2–3 (1990): 235–239, https://doi.org/10.1016/0022-0981(90)90030-G.

17.   U. K. Le Guin, *The Left Hand of Darkness* (New York: Library of America, 1968).

18.   C. W. Petersen,"Reproductive Behavior, Egg Trading, and Correlates of Male Mating Success in the Simultaneous Hermaphrodite, *Serranus tabacarius*," *Environmental Biology of Fishes* 43 (August 1995): 351–361, https://doi.org/10.1007/BF00001169.

19.   E. A. Fischer, "The Relationship Between Mating System and Simultaneous Hermaphroditism in the Coral Reef Fish" *Hypoplectrus nigricans* (Serranidae)," *Animal Behaviour* 28 (May 1980): 620–633, https://doi.org/10.1016/S0003-3472(80)80070-4.

5. ONE GENOME, TWO BODIES

1.   If it is true that Anne Boleyn was executed by England's Henry VIII for not producing a male heir, it is thus extra tragic: that her one surviving child, the future Elizabeth I, was female was actually the king's "fault."

2.   T. K. Solomon-Lane, E. J. Crespi, and M. S. Grober, "Stress and Serial Adult Metamorphosis: Multiple Roles for the Stress Axis in Socially Regulated Sex Change," *Frontiers in Neuroscience* 7 (2013): 210, https://doi.org/10.3389/fnins.2013.00210; E. V. Todd, O. Ortega-Recalde, H. Liu, M. S. Lamm, K. M. Rutherford, H. Cross, M. A. Black, et al., "Stress, Novel Sex Genes, and Epigenetic Reprogramming Orchestrate Socially Controlled Sex Change," *Science Advances* 5, no. 7 (July 2019): eaaw7006, https://doi.org/10.1126/sciadv.aaw7006.

3.   S. G. Brush, "Nettie M. Stevens and the Discovery of Sex Determination by Chromosomes," *Isis* 69, no. 2 (June 1978): 162–172, https://www.jstor.org/stable/230427.

4.   T. H. Morgan, "Recent Theories in Regard to the Determination of Sex," *Popular Science Monthly* 64 (1903): 97–116.

5.   N. M. Stevens, *Studies in Spermatogenesis with Especial Reference to the "Accessory Chromosome"* (Washington, DC: Carnegie Institution of Washington, 1905).

6.   D. E. Kenney and G. G. Borisy, "Thomas Hunt Morgan at the Marine Biological Laboratory: Naturalist and Experimentalist," *Genetics* 181, no. 3 (March 2009): 841–846, https://doi.org/10.1534/genetics.109.101659.

7.   B. Ganetzky and R. S. Hawley, "The Centenary of Genetics: Bridges to the Future," *Genetics* 202, no. 1 (January 2016): 15–23, https://doi.org/10.1534/genetics.115.180182.

8.   T. H. Morgan, "Sex Limited Inheritance in *Drosophila*," *Science* 32, no. 812 (July 1910): 120–122, https://doi.org/10.1126/science.32.812.120.

9.   C. B. Bridges, "Non-Disjunction as Proof of the Chromosome Theory of Heredity," *Genetics* 1, no. 1 (January 1916): 1–52, https://doi.org/10.1093/genetics/1.1.1.

10.   A. H. Sturtevant, "The Linear Arrangement of Six Sex-Linked Factors in *Drosophila*, as Shown by Their Mode of Association," *Journal of Experimental Zoology* 14 (January 2013): 43–59, https://doi.org/10.1002/jez.1400140104.

11.   T. W. Cline, "A Male-Specific Lethal Mutation in *Drosophila melanogaster* That Transforms Sex," *Developmental Biology* 72, no. 2 (October 1979): 266–275, https://doi.org/10.1016/0012-1606(79)90117-9.

12.   Painter would go on to be the school's thirteenth president, from 1946 to 1952, as well as the named defendant in the Supreme Court case *Sweatt v. Painter*, which forced the last school at the University of Texas to admit its first Black student, Heman Marion Sweatt.

13.   If you ever wondered why current research involving humans requires informed consent and oversight by an institutional review board, now you know.

14.   T. S. Painter, "The Y-Chromosome in Mammals," *Science* 53, no. 1378 (May 1921): 503–504, https://doi.org/10.1126/science.53.1378.503

15.   R. Schofield, "Inheritance of Webbed Toes," *Journal of Heredity* 12, no. 9 (November 1921): 400–401.

16.   W. E. Castle, "The Y-Chromosome Type of Sex-Linked Inheritance in Man," *Science* 55, no. 1435 (1922): 703–704.

17.   C. Stern, "The Problem of Complete Y-Linkage in Man," *American Journal of Human Genetics* 9, no. 3 (September 1957): 147–166.

18.   The mix of male and female traits exhibited by XXY men is termed Kleinfelter's Syndrome, after the doctor who first described it.

19.   P. A. Jacobs and J. A. Strong, "A Case of Human Intersexuality Having a Possible Xxy Sex-Determining Mechanism," *Nature* 183, no. 4657 (January 31 1959): 302–303, https://doi.org/10.1038/183302a0

20.   C. E. Ford, K. W. Jones, P. E. Polani, J. C. de Almeida, and J. H. Briggs, "A Sex-Chromosome Anomaly in a Case of Gonadal Dysgenesis (Turner's Syndrome)," *Lancet* 1, no. 7075 (April 1959): 711–713, https://doi.org/10.1016/s0140-6736(59)91893-8.

21.   W. J. Welshons and L. B. Russell, "The Y-Chromosome as the Bearer of Male Determining Factors in the Mouse," *Proceedings of the National Academy of Sciences of the United States of America* 45, no. 4 (April 1959): 560–566, https://doi.org/10.1073/pnas.45.4.560.

22.   A. Delachapelle, H. Hortling, M. Niemi, and J. Wennstroem, "XX Sex Chromosomes in a Human Male: First Case," *Acta Medica Scandinavica* 175, Suppl. 412 (1964): 25–28, https://doi.org/10.1111/j.0954-6820.1964.tb04630.x; C. M. Nagamine, "Sex Reversal in

Mammals," *Advances in Genome Biology* 4 (1996): 53–118, https://doi.org/10.1016/S1067 -5701(96)80005-2.

23. B. M. Cattanach, C. E. Pollard, and S. G. Hawker, "Sex-Reversed Mice: XX and XO Males," *Cytogenetics* 10, no. 5 (1971): 318–337, https://doi.org/10.1159/000130151; L. Singh and K. W. Jones, "Sex Reversal in the Mouse (Mus Musculus) Is Caused by a Recurrent Nonreciprocal Crossover Involving the X and an Aberrant Y Chromosome," *Cell* 28, no. 2 (February 1982): 205–216, https://doi.org/10.1016/0092-8674(82)90338-5. https://www.ncbi.nlm.nih .gov/pubmed/7060127.

24. D. C. Page, R. Mosher, E. M. Simpson, E. M. Fisher, G. Mardon, J. Pollack, B. McGillivray, A. de la Chapelle, and L. G. Brown, "The Sex-Determining Region of the Human Y Chromosome Encodes a Finger Protein," *Cell* 51, no. 6 (December 1987): 1091–1104, https://doi.org/10.1016/0092-8674(87)90595-2.

25. A. H. Sinclair, P. Berta, M. S. Palmer, J. R. Hawkins, B. L. Griffiths, M. J. Smith, J. W. Foster, et al., "A Gene from the Human Sex-Determining Region Encodes a Protein with Homology to a Conserved DNA-Binding Motif," *Nature* 346, no. 6281 (July 1990): 240–244, https://doi.org/10.1038/346240a0; D. C. Page, E. M. Fisher, B. McGillivray, and L. G. Brown, "Additional Deletion in Sex-Determining Region of Human Y Chromosome Resolves Paradox of X,T(Y;22) Female," *Nature* 346, no. 6281 (July 1990): 279–281, https://doi.org/10.1038/346279a0.

26. J. Hodgkin, "One Lucky XX Male: Isolation of the First *Caenorhabditis elegans* Sex-Determination Mutants," *Genetics* 162, no. 4 (December 2002): 1501–1504, https://doi.org /10.1093/genetics/162.4.1501.

27. Hodgkin, "One Lucky XX Male."

28. *C. elegans* genes are stodgily named with a three-letter abbreviation plus a number—far less whimsical than the anything-goes world of fruit fly genetics. But some took it as a haiku-like challenge. For example, a mutation that prevents males from finding their mate's genital opening, *location of vulva-1*, is, of course, abbreviated *lov-1*. A gene's name is in lowercase italics, while the protein it encodes is written in capitals.

29. Its identification was challenging for two reasons. First, *xol-1* mutants only show a defect if they are XO, but most *C. elegans* cultures are completely XX. Second, the XO mutants that are affected die as early embryos, so they are not easy to notice (hence the gene's full name, *XO-lethal-1*). Reminiscent of *Sex-lethal* in flies (described in "The Fly" section in this chapter), this is caused by inappropriate implementation of dosage compensation.

30. C. S. Raymond, C. E. Shamu, M. M. Shen, K. J. Seifert, B. Hirsch, J. Hodgkin, and D. Zarkower, "Evidence for Evolutionary Conservation of Sex-Determining Genes," *Nature* 391 (February 1998): 873–881, https://doi.org/10.1038/35618.

31. Consistent with this, genes encoding DM family proteins have been found in sponges. See Ana Riesgo, Nathan Farrar, Pamela J. Windsor, Gonzalo Giribet, and Sally P. Leys, "The Analysis of Eight Transcriptomes from All Poriferan Classes Reveals Surprising Genetic Complexity in Sponges," *Molecular Biology and Evolution* 31, no. 5 (2014): 1102–20, https://doi.org/10.1093/molbev/msu057.

32. R. E. Lindeman, M. D. Gearhart, A. Minkina, A. D. Krentz, V. J. Bardwell, and D. Zarkower, "Sexual Cell-Fate Reprogramming in the Ovary by Dmrt1," *Current Biology* 25, no. 6 (March 2015): 764–771, https://doi.org/10.1016/j.cub.2015.01.034; C. K. Matson,

M. W. Murphy, A. L. Sarver, M. D. Griswold, V. J. Bardwell, and D. Zarkower, "Dmrt1 Prevents Female Reprogramming in the Postnatal Mammalian Testis," *Nature* 476, no. 7358 (July 2011): 101–104, https://doi.org/10.1038/nature10239.

6. GOING IT ALONE

1. Reported by K. Mettler, "Anna the Anaconda Got Pregnant All by Herself—by 'Virgin Birth,' " *Washington Post*, May 24, 2019. As we see in chapter 8, some snakes share live birth and pregnancy with mammals, through a separate origin.

2. J. Gutekunst, R. Andriantsoa, C. Falckenhayn, K. Hanna, W. Stein, J. Rasamy, and F. Lyko, "Clonal Genome Evolution and Rapid Invasive Spread of the Marbled Crayfish," *Nature Ecology & Evolution* 2, no. 3 (March 2018): 567–573, https://doi.org/10.1038/s41559 -018-0467-9; S. Gibbens, "Growing Population of Crayfish Has One Female Ancestor," *National Geographic News*, February 6, 2018, https://www.nationalgeographic.com/news /2018/02/marbled-crayfish-marmorkrebs-evolution-genes-tumors-spd/.

3. B. Igic, L. Bohs, and J. R. Kohn, "Ancient Polymorphism Reveals Unidirectional Breeding System Shifts," *Proceedings of the National Academy of Sciences of the United States of America* 103, no. 5 (January 2006): 1359–1363, https://doi.org/10.1073/pnas.0506283103.

4. P. Jarne and J. R. Auld, "Animals Mix It up Too: The Distribution of Self-Fertilization among Hermaphroditic Animals," *Evolution* 60, no. 9 (September 2006): 1816–1824, https:// doi.org/10.1554/06-246.1.

5. R. W. Harrington, "Environmentally Controlled Induction of Primary Male Gonochorists from Eggs of the Self-Fertilizing Hermaphroditic Fish, *Rivulus marmoratus* Poey," *Biological Bulletin* 131 (1967): 174–199, https://doi.org/10.2307/1539887; R. W. Harrington, "How Ecological and Genetic Factors Interact to Determine When Self-Fertilizing Hermaphrodites of *Rivulus marmoratus* Change into Functional Secondary Males, with a Reappraisal of the Modes of Intersexuality Among Fishes," *Copeia* 1971, no. 3 (1971): 389–432, https://doi.org/10.2307/1442438.

6. M. Mackiewicz, A. Tatarenkov, A. Perry, J. R. Martin, J. F. Elder Jr., D. L. Bechler, and J. C. Avise, "Microsatellite Documentation of Male-Mediated Outcrossing between Inbred Laboratory Strains of the Self-Fertilizing Mangrove Killifish (*Kryptolebias marmoratus*)," *Journal of Heredity* 97, no. 5 (September–October 2006): 508–513, https://doi .org/10.1093/jhered/esl017; M. Mackiewicz, A. Tatarenkov, D. S. Taylor, B. J. Turner, and J. C. Avise, "Extensive Outcrossing and Androdioecy in a Vertebrate Species That Otherwise Reproduces as a Self-Fertilizing Hermaphrodite," *Proceedings of the National Academy of Sciences of the United States of America* 103, no. 26 (June 2006): 9924–9928, https:// doi.org/10.1073/pnas.0603847103.

7. V. Katju, E. M. LaBeau, K. J. Lipinski, and U. Bergthorsson, "Sex Change by Gene Conversion in a *Caenorhabditis elegans fog-2* Mutant," *Genetics* 180, no. 1 (September 2008): 669–672, https://doi.org/10.1534/genetics.108.090035.

8. K. D. Kallman and R. W. Harrington Jr., "Evidence for the Existence of Homozygous Clones in the Self-Fertilizing Hermaphroditic Teleost *Rivulus marmoratus* (Poey)," *Biological Bulletin* 126 (1964): 101–114, https://doi.org/10.2307/1539420.

9. In American football, this is an exuberant, and often silly, celebratory display that a player may perform after scoring a touchdown, which occurs when they catch or run with the ball in the area at the end of the playing field that is their team's goal.

10. This latter, self-sterile, cross-fertile condition is found naturally in most of *C. elegans'* closest relatives, indicating that hermaphrodites are new inventions.

11. L. T. Morran, M. D. Parmenter, and P. C. Phillips, "Mutation Load and Rapid Adaptation Favour Outcrossing over Self-Fertilization," *Nature* 462, no. 7271 (November 2009): 350–352, https://doi.org/10.1038/nature08496.

12. L. T. Morran, O. G. Schmidt, I. A. Gelarden, R. C. Parrish II, and C. M. Lively, "Running with the Red Queen: Host-Parasite Coevolution Selects for Biparental Sex," *Science* 333, no. 6039 (July 2011): 216–218, https://doi.org/10.1126/science.1206360.

13. E. E. Goldberg, J. R. Kohn, R. Lande, K. A. Robertson, S. A. Smith, and B. Igic, "Species Selection Maintains Self-Incompatibility," *Science* 330, no. 6003 (October 2010): 493–495, https://doi.org/10.1126/science.1194513.

14. E. S. Haag, J. Helder, P. J. Mooijman, D. Yin, and S. Hu, "The Evolution of Uniparental Reproduction in Rhabditina Nematodes: Phylogenetic Patterns, Developmental Causes, and Surprising Consequences," in *Transitions Between Sexual Systems*, ed. J. L. Leonard (Cham, Switzerland: Spring Nature Switzerland AG, 2019), 99–122.

15. M. E. Nasrallah, P. Liu, and J. B. Nasrallah, "Generation of Self-Incompatible *Arabidopsis thaliana* by Transfer of Two S Locus Genes from *A. lyrata*," *Science* 297, no. 5579 (July 2002): 247–9, https://doi.org/10.1126/science.1072205; D. Yin, E. M. Schwarz, C. G. Thomas, R. L. Felde, I. F. Korf, A. D. Cutter, C. M. Schartner, et al., "Rapid Genome Shrinkage in a Self-Fertile Nematode Reveals Sperm Competition Proteins," *Science* 359, no. 6371 (January 2018): 55–61, https://doi.org/10.1126/science.aao0827.

16. E. A. Gladyshev, M. Meselson, and I. R. Arkhipova, "Massive Horizontal Gene Transfer in Bdelloid Rotifers," *Science* 320, no. 5880 (May 2008): 1210–1213, https://doi.org/10.1126/science.1156407.

17. C. Boschetti, A. Carr, A. Crisp, I. Eyres, Y. Wang-Koh, E. Lubzens, T. G. Barraclough, G. Micklem, and A. Tunnacliffe, "Biochemical Diversification through Foreign Gene Expression in Bdelloid Rotifers," *PLoS Genetics* 8, no. 11 (2012): e1003035, https://doi.org/10.1371/journal.pgen.1003035.

18. J. F. Flot, B. Hespeels, X. Li, B. Noel, I. Arkhipova, E. G. Danchin, A Hejnol, et al., "Genomic Evidence for Ameiotic Evolution in the Bdelloid Rotifer *Adineta vaga*," *Nature* 500, no. 7463 (August 2013): 453–457, https://doi.org/10.1038/nature12326.

19. O. A. Vakhrusheva, E. A. Mnatsakanova, Y. R. Galimov, T. V. Neretina, E. S. Gerasimov, S. A. Naumenko, S. G. Ozerova, et al., "Genomic Signatures of Recombination in a Natural Population of the Bdelloid Rotifer *Adineta vaga*," *Nature Communications* 11, no. 1 (December 2020): 6421, https://doi.org/10.1038/s41467-020-19614-y.

20. V. N. Laine, T. B. Sackton, and M. Meselson, "Genomic Signature of Sexual Reproduction in the Bdelloid Rotifer *Macrotrachella quadricornifera*," *Genetics* 220, no. 2 (February 2022), https://doi.org/10.1093/genetics/iyab221

21. M. Parisi, R. Nuttall, D. Naiman, G. Bouffard, J. Malley, J. Andrews, S. Eastman, and B. Oliver, "Paucity of Genes on the *Drosophila* X Chromosome Showing Male-Biased Expression," *Science* 299, no. 5607 (January 2003): 697–700, https://doi.org/10.1126/science.1079190.

22. V. Reinke, I. S. Gil, S. Ward, and K. Kazmer, "Genome-Wide Germline-Enriched and Sex-Biased Expression Profiles in *Caenorhabditis elegans*," *Development* 131, no. 2 (January 2004): 311–323, https://doi.org/10.1242/dev.00914.

23.   C. G. Thomas, R. Li, H. E. Smith, G. C. Woodruff, B. Oliver, and E. S. Haag, "Simplification and Desexualization of Gene Expression in Self-Fertile Nematodes," *Current Biology* 22, no. 22 (November 2012): 2167–72, https://doi.org/10.1016/j.cub.2012.09.038

24.   M. Gershoni and S. Pietrokovski, "The Landscape of Sex-Differential Transcriptome and Its Consequent Selection in Human Adults," *BMC Biology* 15, no. 1 (February 2017): 7, https://doi.org/10.1186/s12915-017-0352-z.

25.   L. R. Garcia, B. LeBoeuf, and P. Koo, "Diversity in Mating Behavior of Hermaphroditic and Male-Female *Caenorhabditis* Nematodes," *Genetics* 175, no. 4 (April 2007): 1761–71, https://doi.org/10.1534/genetics.106.068304; C. J. van der Kooi and T. Schwander, "On the Fate of Sexual Traits under Asexuality," *Biological Reviews Cambridge Philosophical Society* 89, no. 4 (November 2014): 805–819, https://doi.org/10.1111/brv.12078.

26.   D. Yin, E. M. Schwarz, C. G. Thomas, R. L. Felde, I. F. Korf, A. D. Cutter, C. M. Schartner, et al., "Rapid Genome Shrinkage in a Self-Fertile Nematode Reveals Sperm Competition Proteins," *Science* 359, no. 6371 (January 2018): 55–61, https://doi.org/10.1126/science.aao0827.

27.   D. Yin and E. S. Haag, "Evolution of Sex Ratio Through Gene Loss," *Proceedings of the National Academy of Sciences of the United States of America* 116, no. 26 (June 2019): 12919–12924, https://doi.org/10.1073/pnas.1903925116.

28.   See, for example, B. Sykes, *Adam's Curse: A Future Without Men* (New York: Norton, 2004).

7. LAND HO!

1.   T. Deines, "Why Snakes Have Two Penises and Alligators Are Always Erect," *National Geographic News*, February 9, 2018, https://www.nationalgeographic.com/news/2018/02/snakes-alligators-reptiles-genitalia-animals/.

2.   Like mammals, female snakes and lizards have clitorises. These develop from the same paired patches of embryonic tissue as the penis (more on that in chapter 10). Just like these penises, they remain separated to form a pair of *hemiclitores*. Snake clitorises haven't been studied much, but their job is probably to provide sensory input during mating.

3.   A. F. Dixson, *Primate Sexuality: Comparative Studies of the Prosimians, Monkeys, Apes, and Human Beings* (Oxford: Oxford University Press, 1998).

4.   As we will see in the second part of the book, this is part of a larger set of changes that expanded the nonreproductive role of sexual activity in the lineage leading to humans.

5.   C. Y. McLean, P. L. Reno, A. A. Pollen, A. I. Bassan, T. D. Capellini, C. Guenther, V. B. Indjeian, et al., "Human-Specific Loss of Regulatory DNA and the Evolution of Human-Specific Traits," *Nature* 471, no. 7337 (March 2011): 216–219, https://doi.org/10.1038/nature09774.

6.   A. M. Herrera, S. G. Shuster, C. L. Perriton, and M. J. Cohn, "Developmental Basis of Phallus Reduction During Bird Evolution," *Current Biology* 23, no. 12 (June 2013): 1065–1074, https://doi.org/10.1016/j.cub.2013.04.062.

7.   P. L. Brennan, C. J. Clark, and R. O. Prum, "Explosive Eversion and Functional Morphology of the Duck Penis Supports Sexual Conflict in Waterfowl Genitalia," *Proceedings of the Royal Society B* 277, no. 1686 (May 2010): 1309–14, https://doi.org/10.1098/rspb.2009.2139.

8. G. A. Parker, "Sperm Competition and Its Evolutionary Consequences in the Insects," *Biological Reviews* 45 (November 1970): 525–567, https://doi.org/10.1111/j.1469-185X.1970.tb01176.x

9. Smithsonian biologist William Eberhard wrote it: *Sexual Selection and Animal Genitalia* (Princeton, NJ: Princeton University Press, 1985).

10. P. A. Gowaty, "Rape, Forced and Aggressively Coerced Copulation," in *Encyclopedia of Animal Behavior*, 2nd ed., ed. J. C. Choe (New York: Academic Press, 2019), 4: 447–452.

11. M. Panheleux, M. Bain, M. S. Fernandez, I. Morales, J. Gautron, J. L. Arias, S. E. Solomon, M. Hincke, and Y. Nys. "Organic Matrix Composition and Ultrastructure of Eggshell: A Comparative Study," *British Poultry Science* 40, no. 2 (May 1999): 240–252, https://doi.org/10.1080/00071669987665

12. L. D'Alba, D. N. Jones, H. T. Badawy, C. M. Eliason, and M. D. Shawkey, "Antimicrobial Properties of a Nanostructured Eggshell from a Compost-Nesting Bird." *Journal of Experimental Biology* 217, part 7 (April 2014): 1116–21, https://doi.org/10.1242/jeb.098343; L. D'Alba, R. Maia, M. E. Hauber, and M. D. Shawkey, "The Evolution of Eggshell Cuticle in Relation to Nesting Ecology," *Proceedings of the Royal Society B* 283, no. 1836 (August 2016), https://doi.org/10.1098/rspb.2016.0687.

13. K. Stein, E. Prondvai, T. Huang, J. M. Baele, P. M. Sander, and R. Reisz, "Structure and Evolutionary Implications of the Earliest (Sinemurian, Early Jurassic) Dinosaur Eggs and Eggshells," *Scientific Reports* 9, no. 1 (March 2019): 4424, https://doi.org/10.1038/s41598-019-40604-8.

14. G. Ritchison, "Avian Reproduction: Clutch Size, Incubation, and Hatching," Avian Biology (course website), http://avesbiology.com/avianreproduction2.html.

15. R. L. Hughes and L. S. Hall, "Early Development and Embryology of the Platypus," *Philosophical Transactions of the Royal Society (London) B: Biological Sciences* 353, no. 1372 (July 1998): 1101–1114, https://doi.org/10.1098/rstb.1998.0269

16. Beloved in Puerto Rico, that is. The coqui was introduced into the Hawaiian Islands, which historically has no native terrestrial amphibians. It is now regarded as a major threat to native wildlife. See Hawaii Invasive Species Council, "Invasive Species Profiles: Coqui," https://dlnr.hawaii.gov/hisc/info/invasive-species-profiles/coqui/.

8. THE BUN IN THE OVEN

1. T. W. Cranford, M. Amundin, and K. S. Norris, "Functional Morphology and Homology in the Odontocete Nasal Complex: Implications for Sound Generation," *Journal of Morphology* 228, no. 3 (June 1996): 223–285, https://doi.org/10.1002/(SICI)1097-4687(199606)228:3<223::AID-JMOR1>3.0.CO;2-3.

2. D. M. Grossnickle, S. M. Smith, and G. P. Wilson, "Untangling the Multiple Ecological Radiations of Early Mammals," *Trends in Ecology & Evolution* 34, no. 10 (October 2019): 936–949, https://doi.org/10.1016/j.tree.2019.05.008

3. Z. Kielan-Jaworowska, "Pelvic Structure and Nature of Reproduction in Multituberculata," *Nature* 277, no. 5695 (February 1979): 402–403, https://doi.org/10.1038/277402a0.

4. L. N. Weaver, H. Z. Fulghum, D. M. Grossnickle, W. H. Brightly, Z. T. Kulik, G. P. Wilson Mantilla, and M. R. Whitney, "Multituberculate Mammals Show Evidence of a Life

History Strategy Similar to That of Placentals, Not Marsupials," *American Naturalist* 200, no. 3 (September 2022): 383–400, https://doi.org/10.1086/720410.

5. C. X. Yuan, Q. Ji, Q. J. Meng, A. R. Tabrum, and Z. X. Luo, "Earliest Evolution of Multituberculate Mammals Revealed by a New Jurassic Fossil," *Science* 341, no. 6147 (August 2013): 779–783, https://doi.org/10.1126/science.1237970.

6. Q. J. Meng, Q. Ji, Y. G. Zhang, D. Liu, D. M. Grossnickle, and Z. X. Luo, "Mammalian Evolution: An Arboreal Docodont from the Jurassic and Mammaliaform Ecological Diversification," *Science* 347, no. 6223 (February 2015): 764–768, https://doi.org/10.1126/science.1260879; Z. X. Luo, C. X. Yuan, Q. J. Meng, and Q. Ji, "A Jurassic Eutherian Mammal and Divergence of Marsupials and Placentals," *Nature* 476, no. 7361 (August 2011): 442–445, https://doi.org/10.1038/nature10291.

7. The researchers involved in this early work overcame much to make their observations. Several merit a biography, but I allow here only a short digression on two of them. In 1941, the Russian zoologist Alexei Sergeev had just completed his landmark dissertation research on squamate reproduction when he was sent to the front to fight the Germans. Having survived battle and capture in Ukraine, he was "rescued" by his comrades from a German prisoner-of-war (POW) camp. Such freed POWs were treated with suspicion by Soviet authorities. Instead of returning to science after his service, Sergeev was sent to a gulag, where he died in 1943. Equally noteworthy was Australian Claire Weeks, who was the first woman to earn a doctoral degree from the University of Sydney. Having overcome the unrepentant sexism of her time to do important science, she later spent two years as a professional singer in Europe and became a household name through her popular books on the treatment of anxiety. For more, see R. Shine, "Evolution of an Evolutionary Hypothesis: A History of Changing Ideas About the Adaptive Significance of Viviparity in Reptiles," *Journal of Herpetology* 48 no.2 (2014): 147–161, https://doi.org/10.1670/13-075

8. R. Shine, "Reptilian Viviparity in Cold Climates: Testing the Assumptions of an Evolutionary Hypothesis," *Oecologia* 57, no. 3 (March 1983): 397–405, https://doi.org/10.1007/BF00377186.

9. R. Shine, "A New Hypothesis for the Evolution of Viviparity in Reptiles," *American Naturalist* 145, no. 5 (May 1995): 809–823, https://doi.org/10.1086/285769.

10. W. L. Hodges, "Evolution of Viviparity in Horned Lizards (Phrynosoma): Testing the Cold-Climate Hypothesis," *Journal of Evolutionary Biology* 17, no. 6 (November 2004): 1230–1237, https://doi.org/10.1111/j.1420-9101.2004.00770.x.

11. S. J. Gould, "Dollo on Dollo's Law: Irreversibility and the Status of Evolutionary Laws," *Journal of the History of Biology* 3 (Fall 1970): 189–212, https://doi.org/10.1007/BF00137351.

12. I spent my graduate career working on *Heliocidaris erythrogramma*, an Australian sea urchin that made the transition to direct development.

13. C. R. Marshall, E. C. Raff, and R. A. Raff, "Dollo's Law and the Death and Resurrection of Genes," *Proceedings of the National Academy of Sciences of the United States of America* 91, no. 25 (December 1994): 12283–12287, https://doi.org/10.1073/pnas.91.25.12283.

14. K. Smith, "Placental Evolution in Therian Mammals," in *Great Transformations in Vertebrate Evolution*, ed. K. P. Dial, N. Shubin, and E. L. Brainerd (Chicago: University of Chicago Press, 2015), 205–226; S. F. Gilbert, *Developmental Biology*, 6th ed. (Sunderland MA; Sinauer, 2000); C. Ross and T. E. Boroviak, "Origin and Function of the Yolk Sac

in Primate Embryogenesis," *Nature Communications* 11, no. 1 (July 2020): 3760, https://doi .org/10.1038/s41467-020-17575-w.

15. J. G. Schurman, *The Ethical Import of Darwinism*, 2nd ed. (New York: Charles Scribner's Sons 1893).

16. Among the many evo-devo books, these approachable volumes nicely chart the field's growth: R. A. Raff, *The Shape of Life* (Chicago: University of Chicago Press, 1996); S. B. Carroll, *Endless Forms Most Beautiful* (New York: Norton, 2005); A. Wagner, *The Arrival of the Fittest: How Nature Innovates* (New York: Penguin Random House, 2014).

17. X. Tian, J. Gautron, P. Monget, and G. Pascal, "What Makes an Egg Unique? Clues from Evolutionary Scenarios of Egg-Specific Genes," *Biology of Reproduction* 83, no. 6 (December 2010): 893–900, https://doi.org/10.1095/biolreprod.110.085019.

18. Among the many, see T. Werner, S. Koshikawa, T. M. Williams, and S. B. Carroll, "Generation of a Novel Wing Colour Pattern by the Wingless Morphogen," *Nature* 464, no. 7292 (April 2010): 1143–1148, https://doi.org/10.1038/nature08896

19. L. Yadgary, R. Yair, and Z. Uni, "The Chick Embryo Yolk Sac Membrane Expresses Nutrient Transporter and Digestive Enzyme Genes," *Poultry Science* 90, no. 2 (February 2011): 410–416, https://doi.org/10.3382/ps.2010-01075.

20. C. S. Kovacs and H. M. Kronenberg, "Maternal-Fetal Calcium and Bone Metabolism During Pregnancy, Puerperium, and Lactation," *Endocrine Reviews* 18, no. 6 (December 1997): 832–872, https://doi.org/10.1210/edrv.18.6.0319.

21. This explains why expectant mothers are encouraged to take calcium and vitamin supplements.

22. W. G. Breed, "Egg Maturation and Fertilization in Marsupials," *Reproduction, Fertility and Development* 8, no. 4 (1996): 617–643, https://doi.org/10.1071/rd9960617.

23. D. Brawand, W. Wahli, and H. Kaessmann, "Loss of Egg Yolk Genes in Mammals and the Origin of Lactation and Placentation," *PLoS Biology* 6, no. 3 (March 2008): e63, https:// doi.org/10.1371/journal.pbio.0060063.

24. If this were true, the platypus genome (sequenced in 2008) ought to still have one. My own recent scrutiny of that ended in ambiguity: there are genes with a similar sequence, but no information about whether they are active in eggs.

25. K. Smith, "Evolution of the Placenta in Therian Mammals"; J. Spurway, P. Logan, and S. Pak, "The Development, Structure, and Blood Flow within the Umbilical Cord with Particular Reference to the Venous System," *Australasian Journal of Ultrasound in Medicine* 15, no. 3 (August 2012): 97–102, https://doi.org/10.1002/j.2205-0140.2012.tb00013.x; T. Cindrova-Davies, E. Jauniaux, M. G. Elliot, S. Gong, G. J. Burton, and D. S. Charnock-Jones, "RNA-Seq Reveals Conservation of Function Among the Yolk Sacs of Human, Mouse, and Chicken," *Proceedings of the National Academy of Sciences of the United States of America* 114, no. 24 (June 2017): E4753–E4761, https://doi.org/10.1073/pnas.1702560114.

26. A. R. Chavan, O. W. Griffith, and G. P. Wagner, "The Inflammation Paradox in the Evolution of Mammalian Pregnancy: Turning a Foe into a Friend," *Current Opinion in Genetics & Development* 47 (December 2017): 24–32, https://doi.org/10.1016/j.gde.2017.08.004.

27. O. W. Griffith, A. R. Chavan, S. Protopapas, J. Maziarz, R. Romero, and G. P. Wagner, "Embryo Implantation Evolved from an Ancestral Inflammatory Attachment Reaction," *Proceedings of the National Academy of Sciences of the United States of America* 114, no. 32 (August 2017): E6566–E6575, https://doi.org/10.1073/pnas.1701129114.

28.   S. Mi, X. Lee, X. Li, G. M. Veldman, H. Finnerty, L. Racie, E. LaVallie, et al., "Syncytin Is a Captive Retroviral Envelope Protein Involved in Human Placental Morphogenesis," *Nature* 403, no. 6771 (February 2000): 785–789, https://doi.org/10.1038/35001608.

29.   K. A. Dunlap, M. Palmarini, M. Varela, R. C. Burghardt, K. Hayashi, J. L. Farmer, and T. E. Spencer. "Endogenous Retroviruses Regulate Periimplantation Placental Growth and Differentiation," *Proceedings of the National Academy of Sciences of the United States of America* 103, no. 39 (September 2006): 14390–14395, https://doi.org/10.1073/pnas.0603836103; A. Dupressoir, G. Marceau, C. Vernochet, L. Benit, C. Kanellopoulos, V. Sapin, and T. Heidmann, "Syncytin-a and Syncytin-B, Two Fusogenic Placenta-Specific Murine Envelope Genes of Retroviral Origin Conserved in Muridae," *Proceedings of the National Academy of Sciences of the United States of America* 102, no. 3 (January 2005): 725–730, https://doi.org/10.1073/pnas.0406509102.

30.   G. Cornelis, M. Funk, C. Vernochet, F. Leal, O. A. Tarazona, G. Meurice, O. Heidmann, et al., "An Endogenous Retroviral Envelope Syncytin and Its Cognate Receptor Identified in the Viviparous Placental Mabuya Lizard," *Proceedings of the National Academy of Sciences of the United States of America* 114, no. 51 (December 2017): E10991–E11000, https://doi.org/10.1073/pnas.1714590114.

31.   N. Yoshizaki, W. Yamaguchi, S. Ito, and C. Katagiri, "On the Hatching Mechanism of Quail Embryos: Participation of Ectodermal Secretions in the Escape of Embryos from the Vitelline Membrane," *Zoological Science* 17, no. 6 (2000): 751–758, https://doi.org/10.2108/zsj.17.751.

32.   M. Nicolas, *Croyances et pratiques populaires turques concernant les naissances* (Paris: Publications Orientalistes de France; 1972), 91–92.

33.   H. M. Mead, *Tikanga Maori: Living by Maori Values* (Wellington, New Zealand: Huia, 2003).

34.   G. Lewis, "Maternal Mortality in the Developing World: Why Do Mothers Really Die?," *Obstetric Medicine* 1, no. 1 (September 2008): 2–6, https://doi.org/10.1258/om.2008.080019.

35.   UN Human Rights Council, *Report of the Office of the United Nations High Commissioner for Human Rights on Preventable Maternal Mortality and Morbidity and Human Rights*, A/HRC/14/39 (April 16, 2010), https://www.ohchr.org/sites/default/files/Documents/Issues/Women/WRGS/Health/ReportMaternalMortality.pdf

36.   E. A. Phipps, R. Thadhani, T. Benzing, and S. A. Karumanchi, "Pre-Eclampsia: Pathogenesis, Novel Diagnostics and Therapies," *Nature Reviews Nephrology* 15, no. 5 (May 2019): 275–289, https://doi.org/10.1038/s41581-019-0119-6.

37.   S. L. Washburn, "Tools and Human Evolution," *Scientific American* 203 (September 1960): 63–75, https://www.ncbi.nlm.nih.gov/pubmed/13843002

38.   A. G. Warrener, K. L. Lewton, H. Pontzer, and D. E. Lieberman, "A Wider Pelvis Does Not Increase Locomotor Cost in Humans, with Implications for the Evolution of Childbirth," *PLoS One* 10, no. 3 (2015): e0118903, https://doi.org/10.1371/journal.pone.0118903.

39.   H. Dunsworth and L. Eccles, "The Evolution of Difficult Childbirth and Helpless Hominin Infants," *Annual Review of Anthropology* 44 (October 2015): 55–69, https://doi.org/10.1146/annurev-anthro-102214-013918.

40.   M. Pavlicev, R. Romero, and P. Mitteroecker, "Evolution of the Human Pelvis and Obstructed Labor: New Explanations of an Old Obstetrical Dilemma," *American Journal*

of *Obstetrics & Gynecology* 222, no. 1 (January 2020): 3–16, https://doi.org/10.1016/j.ajog .2019.06.043; P. Mitteroecker, S. M. Huttegger, B. Fischer, and M. Pavlicev, "Cliff-Edge Model of Obstetric Selection in Humans," *Proceedings of the National Academy of Sciences of the United States of America* 113, no. 51 (December 2016): 14680–14685, https://doi.org /10.1073/pnas.1612410113.

41. E. Stansfield, K. Kumar, P. Mitteroecker, and N. D. S. Grunstra, "Biomechanical Trade-Offs in the Pelvic Floor Constrain the Evolution of the Human Birth Canal," *Proceedings of the National Academy of Sciences of the United States of America* 118, no. 16 (April 2021): e2022159118, https://doi.org/10.1073/pnas.2022159118.

42. F. Grutzner, W. Rens, E. Tsend-Ayush, N. El-Mogharbel, P. C. O'Brien, R. C. Jones, M. A. Ferguson-Smith, and J. A. Marshall Graves, "In the Platypus a Meiotic Chain of Ten Sex Chromosomes Shares Genes with the Bird Z and Mammal X Chromosomes., *Nature* 432, no. 7019 (December 2004): 913–917, https://doi.org/10.1038/nature03021.

43. F. Veyrunes, P. D. Waters, P. Miethke, W. Rens, D. McMillan, A. E. Alsop, F. Grutzner, et al., "Bird-Like Sex Chromosomes of Platypus Imply Recent Origin of Mammal Sex Chromosomes," *Genome Research* 18, no. 6 (June 2008): 965–973, https://doi.org/10.1101 /gr.7101908; M. C. Wallis, P. D. Waters, M. L. Delbridge, P. J. Kirby, A. J. Pask, F. Grutzner, W. Rens, M. A. Ferguson-Smith, and J. A. Graves, "Sex Determination in Platypus and Echidna: Autosomal Location of Sox3 Confirms the Absence of Sry from Monotremes," *Chromosome Research* 15, no. 8 (2007): 949–959, https://doi.org/10.1007/s10577-007-1185-3.

44. D. P. Barlow and M. S. Bartolomei, "Genomic Imprinting in Mammals," *Cold Spring Harbor Perspectives in Biology* 6, no. 2 (February 2014), https://doi.org/10.1101/cshperspect .a018382.

9. PERIOD PIECE

1. J. Thomas, K. Handasyde, M. L. Parrott, and P. Temple-Smith, "The Platypus Nest: Burrow Structure and Nesting Behaviour in Captivity," *Australian Journal of Zoology* 65, no. 6 (2017): 347–356, https://doi.org/https://doi.org/10.1071/ZO18007.

2. M. Pavlicev and G. Wagner, "The Evolutionary Origin of Female Orgasm," *Journal of Experimental Zoology Part B: Molecular and Developmental Evolution* 326, no. 6 (September 2016): 326–337, https://doi.org/10.1002/jez.b.22690.

3. D. Emera, R. Romero, and G. Wagner, "The Evolution of Menstruation: A New Model for Genetic Assimilation: Explaining Molecular Origins of Maternal Responses to Fetal Invasiveness," *Bioessays* 34, no. 1 (January 2012): 26–35, https://doi.org/10.1002/bies.201100099.

4. M. Profet, "Menstruation as a Defense against Pathogens Transported by Sperm," *Quarterly Review of Biology* 68, no. 3 (September 1993): 335–386, https://doi.org/10.1086/418170.

5. B. I. Strassmann, "The Evolution of Endometrial Cycles and Menstruation," *Quarterly Review of Biology* 71, no. 2 (June 1996): 181–220, https://doi.org/10.1086/419369. 8693059

6. G. Teklenburg, M. Salker, M. Molokhia, S. Lavery, G. Trew, T. Aojanepong, H. J. Mardon, et al., "Natural Selection of Human Embryos: Decidualizing Endometrial Stromal Cells Serve as Sensors of Embryo Quality upon Implantation," *PLoS One* 5, no. 4 (April 2010): e10258, https://doi.org/10.1371/journal.pone.0010258.

7. Reviewed in Pavlicev and Wagner, "The Evolutionary Origin of Female Orgasm."

8.   J. Bakker and M. J. Baum. "Neuroendocrine Regulation of Gnrh Release in Induced Ovulators," *Frontiers in Neuroendocrinology* 21, no. 3 (July 2000): 220–262, https://doi.org /10.1006/frne.2000.0198.

9.   M. Pavlicev, A. M. Zupan, A. Barry, S. Walters, K. M. Milano, H. J. Kliman, and G. P. Wagner, "An Experimental Test of the Ovulatory Homolog Model of Female Orgasm," *Proceedings of the National Academy of Sciences of the United States of America* 116, no. 41 (October 2019): 20267–20273, https://doi.org/10.1073/pnas.1910295116.

10.   P. L. Vasey, A. Foroud, N. Duckworth, and S. D. Kovacovsky, "Male-Female and Female-Female Mounting in Japanese Macaques: A Comparative Study of Posture and Movement," *Archives of Sexual Behavior* 35, no. 2 (April 2006): 117–129, https://doi.org /10.1007/s10508-005-9007-1.

11.   K. Wallen and E. A. Lloyd, "Clitoral Variability Compared with Penile Variability Supports Nonadaptation of Female Orgasm," *Evolution & Development* 10, no. 1 (January– February 2008): 1–2, https://doi.org/10.1111/j.1525-142X.2007.00207.x.

12.   S. Naguib, "Horizons and Limitations of Muslim Feminist Hermeneutics: Reflections on the Menstruation Verse," in *New Topics in Feminist Philosophy of Religion*, ed. P. S. Anderson (New York: Springer, 2010), 33–49.

13.   M. Espada, "Period Taboo Around the World," *Teen Vogue Newsletter*, May 28, 2018, https:// www.teenvogue.com/story/period-taboo-around-the-world.

14.   E. Silverman, "Three Virginia Teens Fight 'Period Poverty,' " *Washington Post*, December 10, 2022, https://www.proquest.com/newspapers/three-virginia-teens-fight-period-poverty /docview/2748646349/se-2

15.   U.S. Agency for International Development, *Empowerment of Women Through Sanitation and Hygiene Programs, FY 2022*, USAID report to Congress (Washington, DC: U.S. Government Publishing Office, 2023). https://www.usaid.gov/reports/empowerment -women-sanitation-hygiene/fy-2022

16.   P. Rucker, "Trump Says Fox's Megyn Kelly Had 'Blood Coming Out of Her Wherever,' " *Washington Post*, August 8, 2015, https://www.washingtonpost.com/news/post-politics/wp /2015/08/07/trump-says-foxs-megyn-kelly-had-blood-coming-out-of-her-wherever/.

17.   T. Buckley and A. Gottlieb, eds., *Blood Magic: The Anthropology of Menstruation* (Berkeley: University of California Press, 1988).

18.   M. S. M. Pavelka and L. M. Fedigan, "Menopause: A Comparative Life History Perspective," *American Journal of Physical Anthropology* 34, Suppl. 13, American Journal of Physical Anthropology Annual Meeting issue (1991): 13–38, https://doi.org/10.1002/ajpa.1330340604.

19.   J. Chamie, "Increasingly Indispensable Grandparents," YaleGlobal Online, September 4, 2018, https://archive-yaleglobal.yale.edu/content/increasingly-indispensable-grandparents.

20.   G. C. Williams, "Pleiotropy, Natural Selection, and the Evolution of Senescence," *Evolution* 11, no. 4 (December 1957): 398–411, https://doi.org/10.2307/2406060.

10. SCENTS AND SENSIBILITY

1.   L. Stowers, T. E. Holy, M. Meister, C. Dulac, and G. Koentges, "Loss of Sex Discrimination and Male-Male Aggression in Mice Deficient for Trp2," *Science* 295, no. 5559 (February 2002): 1493–1500, https://doi.org/10.1126/science.1069259.

2.   T. Kimchi, J. Xu, and C. Dulac, "A Functional Circuit Underlying Male Sexual Behaviour in the Female Mouse Brain," *Nature* 448, no. 7157 (August 2007): 1009–1014, https://doi .org/10.1038/nature06089.

3.   K. P. Bhatnagar and E. Meisami, "Vomeronasal Organ in Bats and Primates: Extremes of Structural Variability and Its Phylogenetic Implications," *Microscopy Research and Technique* 43, no. 6 (December 1998): 465–475, https://doi.org/10.1002/(SICI)1097-0029 (19981215)43:6<465::AID-JEMT1>3.0.CO;2-1; T. Kishida, J. Thewissen, T. Hayakawa, H. Imai, and K. Agata, "Aquatic Adaptation and the Evolution of Smell and Taste in Whales," *Zoological Letters* 1 (2015): 9, https://doi.org/10.1186/s40851-014-0002-z; A. Mack-ay-Sim, D. Duvall, and B. M. Graves, "The West Indian Manatee (Trichechus Manatus) Lacks a Vomeronasal Organ," *Brain, Behavior, and Evolution* 27, no. 2–4 (1985): 186–194, https://doi.org/10.1159/000118729; P. Perelman, W. E. Johnson, C. Roos, H. N. Seuanez, J. E. Horvath, M. A. Moreira, B. Kessing, et al., "A Molecular Phylogeny of Living Primates," *PLoS Genetics* 7, no. 3 (March 2011): e1001342, https://doi.org/10.1371/journal .pgen.1001342.

4.   D. Trotier, C. Eloit, M. Wassef, G. Talmain, J. L. Bensimon, K. B. Doving, and J. Ferrand, "The Vomeronasal Cavity in Adult Humans," *Chemical Senses* 25, no. 4 (August 2000): 369–380, https://doi.org/10.1093/chemse/25.4.369.

5.   D. W. Frayer and M. H. Wolpoff, "Sexual Dimorphism," *Annual Review of Anthropology* 14 (1985): 429–473, https://doi.org/10.1146/annurev.an.14.100185.002241.

6.   L. T. Kozlowski and J. T. Cutting, "Recognizing the Sex of a Walker from a Dynamic Point-Light Display," *Perception & Psychophysics*, 21, no. 6 (1977): 575–580, https://doi.org /10.3758/BF03198740.

7.   A. Parkes, V. Strange, D. Wight, C. Bonell, A. Copas, M. Henderson, K. Buston, et al., "Comparison of Teenagers' Early Same-Sex and Heterosexual Behavior: UK Data from the Share and Ripple Studies," *Journal of Adolescent Health* 48, no. 1 (January 2011): 27–35, https://doi.org/10.1016/j.jadohealth.2010.05.010.

8.   A. Kinsey, W. Pomeroy, and C. Martin, *Sexual Behavior in the Human Male* (Philadelphia: Saunders, 1948); A. Kinsey, W. Pomeroy, C. Martin, and P. Gebhard, *Sexual Behavior in the Human Female* (Philadelphia: Saunders, 1953); A. Chandra, W. D. Mosher, C. Copen, and C. Sionean, "Sexual Behavior, Sexual Attraction, and Sexual Identity in the United States: Data from the 2006–2008 National Survey of Family Growth," *National Health Statistics Reports*, no. 36 (March 2011): 1–36, https://www.ncbi.nlm.nih.gov/pubmed/21560887.

9.   For example, see R. Green, *The "Sissy Boy Syndrome" and the Development of Human Homosexuality* (New Haven, CT: Yale University Press, 1987).

10. "Bling" is an urban North American slang term for showy jewelry.

11. E. Trinkaus and A. P. Buzhilova, "The Death and Burial of Sunghir 1," *International Journal of Osteoarchaeology* 22 (2012): 655–666, https://doi.org/10.1002/oa.1227.

11. THE MALE GAZE

1.   I stop short of saying the peacock's tail is actually detrimental to daily function because it appears to be inconsequential for the short escape flights peafowl use to escape predators. For more, see G. N. Askew, "The Elaborate Plumage in Peacocks Is Not Such a

Drag," *Journal of Experimental Biology* 217, pt. 18 (September 2014): 3237–3241, https://doi
.org/10.1242/jeb.107474.

2. A. Pomiankowski, "Sexual Selection: The Handicap Principle Does Work—Sometimes,"
    Proceedings of the Royal Society of London, Series B: Biological Sciences 231 (June 1987): 123–
    145, https://doi.org/10.1098/rspb.1987.0038.

3. T. Janicke, I. K. Haderer, M. J. Lajeunesse, and N. Anthes, "Darwinian Sex Roles Con-
    firmed across the Animal Kingdom," *Science Advances* 2, no. 2 (February 2016): e1500983,
    https://doi.org/10.1126/sciadv.1500983.

4. M. Ah-King and S. Nylin, "Sex in an Evolutionary Perspective: Just Another Reaction
    Norm," *Evolutionary Biology* 37, no. 4 (December 2010): 234–246, https://doi.org/10.1007
    /s11692-010-9101-8; D. Bachtrog, J. E. Mank, C. L. Peichel, M. Kirkpatrick, S. P. Otto, T. L.
    Ashman, M. W. Hahn, et al., "Sex Determination: Why So Many Ways of Doing It?,"
    PLoS Biology 12, no. 7 (July 2014): e1001899, https://doi.org/10.1371/journal.pbio.1001899.

5. K. A. Paczolt and A. G. Jones, "Post-Copulatory Sexual Selection and Sexual Conflict in
    the Evolution of Male Pregnancy," *Nature* 464, no. 7287 (March 2010): 401–404, https://
    doi.org/10.1038/nature08861.

6. A. Berglund and G. Rosenqvist, "Male Pipefish Prefer Ornamented Females," *Animal
    Behavior* 61, no. 2 (February 2001): 345–350, https://doi.org/10.1006/anbe.2000.1599.

7. T. R. Mace, "Time Budget and Pair-Bond Dynamics in the Comb-Crested Jacana *Iredi-
    parra gallinacea*: A Test of Hypotheses," *Emu*, 100, no. 1 (2000): 31–41, https://doi.org/10.1071
    /MU9844; S. T. Emlen and P. H. Wrege, "Size Dimorphism, Intrasexual Competition,
    and Sexual Selection in Wattled Jacana (*Jacana jacana*), a Sex-Role-Reversed Shorebird in
    Panama," *Auk* 121, no. 2 (April 2004): 391–403, https://doi.org/10.1093/auk/121.2.391.

8. I. Schlupp, *Male Choice, Female Competition, and Female Ornaments* (New York: Oxford
    University Press, 2021).

9. J. A. Tobias, R. Montgomerie, and B. E. Lyon, "The Evolution of Female Ornaments and
    Weaponry: Social Selection, Sexual Selection and Ecological Competition," *Philosophical
    Transactions of the Royal Society of London, Series B: Biological Sciences* 367, no. 1600 (August
    2012): 2274–2293, https://doi.org/10.1098/rstb.2011.0280.

10. R. Heinsohn, S. Legge, and J. A. Endler, "Extreme Reversed Sexual Dichromatism in a
    Bird without Sex Role Reversal," *Science* 309, no. 5734 (July 2005): 617–619, https://doi.org
    /10.1126/science.1112774.

11. R. Heinsohn, "Ecology and Evolution of the Enigmatic Eclectus Parrot (*Eclectus rora-
    tus*)," *Journal of Avian Medicine and Surgery* 22, no. 2 (June 2008): 146–150, https://doi.org
    /10.1647/2007-031.1.

12. R. Chan, D. Stuart-Fox, and T. S. Jessop, "Why Are Females Ornamented? A Test of the
    Courtship Stimulation and Courtship Rejection Hypotheses," *Behavioral Ecology* 20, no.
    6 (November 2009): 1334–1342, https://doi.org/10.1093/beheco/arp136.

13. G. G. Watkins, "Inter-Sexual Signalling and the Functions of Female Coloration in the
    Tropidurid Lizard *Microlophus occipitalis*," *Animal Behavior* 53, no. 4 (April 1997): 843–852,
    https://doi.org/10.1006/anbe.1996.0350.

14. A. F. Dixson, "Human Sexual Dimorphism: Opposites Attract," chap. 7 in *Sexual Selection
    and the Origins of Human Mating Systems* (Oxford: Oxford University Press, 2009), 124–154.

15. R. G. Abramson, A. Mavi, T. Cermik, S. Basu, N. E. Wehrli, M. Houseni, S. Mishra, et al., "Age-Related Structural and Functional Changes in the Breast: Multimodality Correlation with Digital Mammography, Computed Tomography, Magnetic Resonance Imaging, and Positron Emission Tomography," *Seminars in Nuclear Medicine* 37, no. 3 (May 2007): 146–153, https://doi.org/10.1053/j.semnuclmed.2007.01.003.

16. Dixson, "Human Sexual Dimorphism."

17. Hooters is an American restaurant chain famous for its large-breasted waitresses in tight shirts. Its corporate logo depicts an owl, and owls do hoot, but "hooter" is also a slang term for breast, and the owl's eyes are suspiciously booby.

18. C. Lévi-Strauss, *Les structures élémentaires d la parenté* (Paris: Presses Universitaires de France, 1947).

19. B. Chapais, *Primeval Kinship: How Pair-Bonding Gave Birth to Human Society* (Cambridge, MA: Harvard University Press, 2008).

20. M. R. Fales, D. A. Frederick, J. R. Garcia, K. A. Gildersleeve, M. G. Haselton, and H. E. Fisher, "Mating Markets and Bargaining Hands: Mate Preferences for Attractiveness and Resources in Two National U.S. Studies," *Personality & Individual Differences* 88 (January 2016): 78–87, http://dx.doi.org/10.1016/j.paid.2015.08.041.

21. K. E. Starkweather and R. Hames, "A Survey of Non-Classical Polyandry," *Human Nature* 23 (2012): 149–172, https://doi.org/10.1007/s12110-012-9144-x.

22. S. Beckerman and P. Valentine, eds., *Cultures of Multiple Fathers: The Theory and Practice of Partible Paternity in Lowland South America* (Gainesville: University of Florida Press, 2002).

23. S. B. Hrdy, "The 'One Animal in All Creation About Which Man Knows the Least,'" *Philosophical Transactions of the Royal Society of London, Series B: Biological Sciences* 368, no. 1631 (2013): 20130072, https://doi.org/10.1098/rstb.2013.0072; Schlupp, *Male Choice, Female Competition.*

24. M. L. Fisher and A. Cox, "Four Strategies Used During Intrasexual Competition for Mates," *Personal Relationships*, 18, no. 1 (January 2011), 20–38, https://doi.org/10.1111/j.1475-6811.2010.01307.x.

25. M. L. Fisher and C. Candea, "You Ain't Woman Enough to Take My Man: Female Intrasexual Competition as Portrayed in Songs," *Journal of Social, Evolutionary, & Cultural Psychology* 6, no. 4 (December 2012): 480–493, https://doi.org/10.1037/h0099238.

26. E. Graham-Harrison, "'I Daren't Go Far': Taliban Rules Trap Afghan Women with No Male Guardian," *The Guardian*, August 15, 2022, https://www.theguardian.com/world/2022/aug/15/taliban-rules-trap-afghan-women-no-male-guardian.

27. Georgetown Institute for Women, Peace, and Security, "Mahram: Women's Mobility in Islam" (2022), https://giwps.georgetown.edu/wp-content/uploads/2022/08/Mahram-Womens-Mobility-in-Islam.pdf.

28. G. Hale, T. Regev, and Y. Rubinstein, "Do Looks Matter for an Academic Career in Economics?" (working paper, CEPR Discussion Papers, no. 15893, Center for Economic Policy Research, London, 2021), https://ideas.repec.org/p/cpr/ceprdp/15893.html.

29. C. Hakim, *Erotic Capital* (New York: Basic Books, 2011).

12. SEX VERSUS SEXISM

1. See, for example, A. Fausto-Sterling, "Why Sex Is Not Binary," *New York Times*, October 29, 2018.

2. Data replotted from A. E. Ellibes Kaya, O. Dogan, M. Yassa, et al., "Do External Female Genital Measurements Affect Genital Perception and Sexual Function and Orgasm?," *Turkish Journal of Obstetrics & Gynecology* 17 (2020): 175–181, https://doi:10.4274/tjod .galenos.2020.89896; D. Veal, S. Miles, S. Bramley, et al., "Am I Normal? A Systematic Review and Construction of Nomograms for Flaccid and Erect Penis Length and Circumference in Up to 15 521 Men," *BJU International* 115 (2015): 978–986, https://doi.org /10.1111/bju.13010.

3. N. Samolovitch, "Brilliant Women in US Horse Racing History," America's Best Racing, March 8, 2022, https://www.americasbestracing.net/the-sport/2020-brilliant-women-us -horse-racing-history.

4. "The History of Women Jockeys in the Kentucky Derby," Derby Experiences, https:// derbyexperiences.com/blog/women-jockeys-in-the-kentucky-derby (accessed February 5, 2021). As of this writing, the post has been taken down.

5. U. Thyen, K. Lanz, P. M. Holterhus, and O. Hiort, "Epidemiology and Initial Management of Ambiguous Genitalia at Birth in Germany," *Hormone Research* 66, no. 4 (2006): 195–203, https://doi.org/10.1159/000094782.

6. J. K. Petersen, *A Comprehensive Guide to Intersex* (Philadelphia: Jessica Kingsley, 2021).

7. C. Fine, *Delusions of Gender: How Our Minds, Society, and Neurosexism Create Difference* (New York: Norton, 2010).

8. B. K. Todd, R. A. Fischer, S. Di Costa, A. Roestorf, K. Harbour, P. Hardiman, and J. A. Barry, "Sex Differences in Children's Toy Preferences: A Systematic Review, Meta-Regression, and Meta-Analysis," *Infant and Child Development* 27, no. 2 (March 2018): e2064, https://doi.org/10.1002/icd.2064.

9. S. A. Berenbaum and M. Hines, "Early Androgens Are Related to Childhood Sex-Typed Toy Preferences," *Psychological Science* 3, no. 3 (2017): 203–206, https://doi.org /10.1111/j.1467-9280.1992.tb00028.x.

10. D. Joel and L. Vikhanski, *Gender Mosaic: Beyond the Myth of the Male and Female Brain* (New York: Little, Brown Spark, 2019).

11. S. Richardson, *Sex Itself: The Search for Male and Female in the Human Genome* (Chicago: University of Chicago Press; 2013).

12. M. L. Barr and E. G. Bertram, "A Morphological Distinction between Neurones of the Male and Female, and the Behaviour of the Nucleolar Satellite During Accelerated Nucleoprotein Synthesis," *Nature* 163, no. 4148 (April 1949): 676, https://doi.org /10.1038/163676a0.

13. H. Fang, C. M. Disteche, and J. B. Berletch, "X Inactivation and Escape: Epigenetic and Structural Features," *Frontiers in Cell Development and Biology* 7 (2019): 219, https://doi .org/10.3389/fcell.2019.00219.

14. H. Meischke, M. P. Larsen, and M. S. Eisenberg, "Gender Differences in Reported Symptoms for Acute Myocardial Infarction: Impact on Prehospital Delay Time Interval," *American Journal of Emergency Medicine* 16 (July 1998): 363–366, https://doi:10.1016 /s0735-6757(98)90128-0.

15. R. V. Clark, J. A. Wald, R. S. Swerdloff, C. Wang, F. C. W. Wu, L. D. Bowers, and A. M. Matsumoto, "Large Divergence in Testosterone Concentrations between Men and Women: Frame of Reference for Elite Athletes in Sex-Specific Competition in Sports, a Narrative Review," *Clinical Endocrinology (Oxf)* 90, no. 1 (January 2019): 15–22, https://doi.org/10.1111/cen.13840.

16. S. M. Rennison, "Rape and Sexual Assault: Reporting to Police and Medical Attention, 1992–2000," *Bureau of Justice Statistics Selected Findings* (Washington, DC: August 2002); here, "rape" is defined as nonconsensual penetration involving a sex organ.

17. United States Sentencing Commission, "Quick Facts: Sexual Abuse Offenders" (2018), https://www.ussc.gov/sites/default/files/pdf/research-and-publications/quick-facts/Sexual_Abuse_FY18.pdf.

18. L. Stemple, A. Flores, and I. H. Meyer, "Sexual Victimization Perpetrated by Women: Federal Data Reveal Surprising Prevalence," *Aggression & Violent Behavior* 34 (May 2017): 302–311, https://doi.org/10.1016/j.avb.2016.09.007.

19. N. Stripe, "Appendix Tables: Homicide in England and Wales" (2020), Office for National Statistics (UK), https://www.ons.gov.uk/peoplepopulationandcommunity/crimeandjustice/datasets/appendixtableshomicideinenglandandwales.

20. Office of Juvenile Justice and Delinquency Prevention, "Easy Access to the FBI's Supplementary Homicide Reports: 1980–2020," Pittsburgh, PA, https://www.ojjdp.gov/ojstatbb/ezashr/.

21. O. Diaz, "Rise in Strangulations Worries Fairfax County," *Washington Post*, March 7, 2023.

22. S. Brownmiller, *Against Our Will: Men, Women, and Rape* (New York: Fawcett, 1975).

13. QUEERLY NORMAL

1. N. G. Jablonski and G. Chaplin, "Colloquium Paper: Human Skin Pigmentation as an Adaptation to UV Radiation," *Proceedings of the National Academy of Sciences of the United States of America* 107, suppl. 2 (May 11, 2010): 8962–8968, https://doi.org/10.1073/pnas.0914628107.

2. J. Tschantret, "Revolutionary Homophobia: Explaining State Repression Against Sexual Minorities," *British Journal of Political Science* (2019): 1–22, https://doi.org/10.1017/S0007123418000480; Office of the United Nations High Commissioner for Human Rights, *Discrimination and Violence Against Individuals Based on Their Sexual Orientation and Gender Identity* (2015), https://digitallibrary.un.org/record/797193?ln=en.

3. C. Tentelier, O. Lepais, N. Larranaga, A. Manicki, F. Lange, and J. Rives, "Sexual Selection Leads to a Tenfold Difference in Reproductive Success of Alternative Reproductive Tactics in Male Atlantic Salmon," *Naturwissenschaften* 103, no. 5–6 (June 2016): 47, https://doi.org/10.1007/s00114-016-1372-1.

4. S. E. Hoover, C. I. Keeling, M. L. Winston, and K. N. Slessor, "The Effect of Queen Pheromones on Worker Honey Bee Ovary Development," *Naturwissenschaften* 90, no. 10 (October 2003): 477–480, https://doi.org/10.1007/s00114-003-0462-z.

5. C. C. Grueter and T. S. Stoinski, "Homosexual Behavior in Female Mountain Gorillas: Reflection of Dominance, Affiliation, Reconciliation or Arousal?," *PLoS One* 11, no. 5 (2016): e0154185, https://doi.org/10.1371/journal.pone.0154185; B. Fruth and G. Hohmann,

"Social Grease for Females? Same-Sex Genital Contacts in Wild Bonobos," in *Homosexual Behaviour in Animals: An Evolutionary Perspective*, ed. V. Sommer and P. L. Vasey (New York: Cambridge University Press, 2006), 294–315.

6.   Although adoption is, of course, a path to parenthood, it is not the evolutionary equivalent to biological parenthood.

7.   A. Chandra, W. D. Mosher, and C. Copen, "Sexual Behavior, Sexual Attraction, and Sexual Identity in the United States: Data from the 2006–2008 National Survey of Family Growth," *National Health Statistics Reports* 36 (March 2011), https://www.cdc.gov/nchs/data/nhsr/nhsr036.pdf.

8.   D. Lofquist, "Same-Sex Couple Households," in *American Community Survey Briefs*, ed. U.S. Census Bureau (Washington, DC: U.S. Department of Commerce, 2011); U.S. Census Bureau, "2015 American Community Survey, 1-Year Estimates" (Washington, DC: U.S. Government Publishing Office, 2015), https://www.census.gov/programs-surveys/acs/technical-documentation/table-and-geography-changes/2015/1-year.html.

9.   A. Kinsey, W. Pomeroy, and C. Martin, *Sexual Behavior in the Human Male* (Philadelphia: Saunders; 1948); A. Kinsey, W. Pomeroy, C. Martin, and P. Gebhard, *Sexual Behavior in the Human Female* (Philadelphia: Saunders; 1953).

10.   L. M. Diamond, "What Develops in the Biodevelopment of Sexual Orientation?," *Archives of Sexual Behavior* (January 25, 2023), https://doi:10.1007/s10508-023-02542-5.

11.   A. P. Moczek and D. D.J. Emlen, "Proximate Determination of Male Horn Dimorphism in the Beetle," *Onthophagus taurus* (Coleoptera: Scarabaeidae)," *Journal of Evolutionary Biology* 12 no. 1 (1999): 27–37, https://doi.org/10.1046/j.1420-9101.1999.00004.x.

12.   D. Ryabko and Z. Reznikova, "On the Evolutionary Origins of Differences in Sexual Preferences," *Frontiers in Psychology* 6 (2015): 628, https://doi.org/10.3389/fpsyg.2015.00628.

13.   Vladziu Valentino ("Lee") Liberace (1919–1987) was a popular American pianist known for his flamboyant dress and manner. Liberace was gay, but straight women formed the core of his fan base.

14.   For reasons having to do with the history of academia and colonial power, cultural anthropology is still dominated by European and North American scholars. "Traditional" thus often means "not wholly of the developed world from which the anthropologists doing the studying mostly come," and therefore needs to be taken with a grain of salt.

15.   D. P. VanderLaan, Z. Ren, and P. L. Vasey, "Male Androphilia in the Ancestral Environment. An Ethnological Analysis," *Human Nature* 24, no. 4 (December 2013): 375–401, https://doi.org/10.1007/s12110-013-9182-z.

16.   A. S. Camperio Ciani, L. Fontanesi, F. Iemmola, E. Giannella, C. Ferron, and L. Lombardi, "Factors Associated with Higher Fecundity in Female Maternal Relatives of Homosexual Men," *Journal of Sexual Medicine* 9, no. 11 (November 2012): 2878–2887, https://doi.org/10.1111/j.1743-6109.2012.02785.x.

17.   J. M. Bailey, M. P. Dunne, and N. G. Martin, "Genetic and Environmental Influences on Sexual Orientation and Its Correlates in an Australian Twin Sample," *Journal of Personality and Social Psychology* 78, no. 3 (March 2000): 524–536, https://doi.org/10.1037/0022-3514.78.3.524.

18.   A. R. Sanders, G. W. Beecham, S. Guo, K. Dawood, G. Rieger, J. A. Badner, E. S. Gershon, et al., "Genome-Wide Association Study of Male Sexual Orientation," *Scientific Reports*

7, no. 1 (December 2017): 16950, https://doi.org/10.1038/s41598-017-15736-4; A. Ganna, K. J. H. Verweij, M. G. Nivard, R. Maier, R. Wedow, A. S. Busch, A. Abdellaoui, et al., "Large-Scale Gwas Reveals Insights into the Genetic Architecture of Same-Sex Sexual Behavior," *Science* 365, no. 6456 (August 2019), https://doi.org/10.1126/science.aat7693.

19.   W. R. Rice, U. Friberg, and S. Gavrilets, "Homosexuality as a Consequence of Epigenetically Canalized Sexual Development," *Quarterly Review of Biology* 87, no. 4 (December 2012): 343–368, https://doi.org/10.1086/668167.

20.   P. Doerr, G. Kockott, H. J. Vogt, K. M. Pirke, and F. Dittmar, "Plasma Testosterone, Estradiol, and Semen Analysis in Male Homosexuals," *Archives of General Psychiatry* 29, no. 6 (December 1973): 829–833, https://doi.org/10.1001/archpsyc.1973.04200060101016.

21.   D. P. Vanderlaan, R. Blanchard, H. Wood, and K. J. Zucker, "Birth Order and Sibling Sex Ratio of Children and Adolescents Referred to a Gender Identity Service," *PLoS One* 9, no. 3 (2014): e90257, https://doi.org/10.1371/journal.pone.0090257; G. Schwartz, R. M. Kim, A. B. Kolundzija, G. Rieger, and A. R. Sanders, "Biodemographic and Physical Correlates of Sexual Orientation in Men," *Archives of Sexual Behavior* 39, no. 1 (February 2010): 93–109, https://doi.org/10.1007/s10508-009-9499-1; R. Blanchard, "Birth Order and Sibling Sex Ratio in Homosexual Versus Heterosexual Males and Females," *Annual Review of Sex Research* 8 (1997): 27–67; R. Blanchard and A. F. Bogaert, "Proportion of Homosexual Men Who Owe Their Sexual Orientation to Fraternal Birth Order: An Estimate Based on Two National Probability Samples," *American Journal of Human Biology* 16, no. 2 (March–April 2004): 151–157, https://doi.org/10.1002/ajhb.20006.

22.   It has been proposed, however, to function potentially through the maternal immune system. See R. Blanchard, "Fraternal Birth Order and the Maternal Immune Hypothesis of Male Homosexuality," *Hormones and Behavior* 40, no. 2 (September 2001): 105–114, https://doi.org/10.1006/hbeh.2001.1681.

23.   R. Green, *The "Sissy Boy Syndrome" and the Development of Homosexuality* (New Haven, CT: Yale University Press; 1987); J. M. Bailey and K. J. Zucker, "Childhood Sex-Typed Behavior and Sexual Orientation: A Conceptual Analysis and Quantitative Review," *Developmental Psychology* 31 no. 1 (1995): 43–55, https://doi.org/10.1037/0012-1649.31.1.43.

24.   P. Richerson, R. Baldini, A. V. Bell, K. Demps, K. Frost, V. Hillis, S. Mathew, et al., "Cultural Group Selection Plays an Essential Role in Explaining Human Cooperation: A Sketch of the Evidence," *Behavioral and Brain Sciences* 39 (January 2016): e30, https://doi.org/10.1017/S0140525X1400106X; D. S. Wilson and E. O. Wilson, "Rethinking the Theoretical Foundation of Sociobiology," *Quarterly Review of Biology* 82, no. 4 (December 2007): 327–348, https://doi.org/10.1086/522809.

25.   D. P. VanderLaan and P. L. Vasey, "Relationship Status and Elevated Avuncularity in Samoan *Fa'afafine*," *Personal Relationships* 19, no. 2 (June 2012): 326–339, https://doi.org/10.1111/j.1475-6811.2011.01364.x; P. L. Vasey and D. P. VanderLaan, "Avuncular Tendencies and the Evolution of Male Androphilia in Samoan Fa'afafine," *Archives of Sexual Behavior* 39, no. 4 (August 2010): 821–830, https://doi.org/10.1007/s10508-008-9404-3.

26.   G. Childs, "Old-Age Security, Religious Celibacy, and Aggregate Fertility in a Tibetan Population," *Journal of Population Research* 18 (2001): 52–67, https://doi.org/10.1007/BF03031955.

27.   D. K. Deady, M. J. Law Smith, J. P. Kent, and R. I. M. Dunbar, "Is Priesthood an Adaptive Strategy?," *Human Nature* 17 (2006): 393–404, https://doi.org/10.1007/s12110-006-1002-2.

28.   N. W. Bailey and M. Zuk, "Same-Sex Sexual Behavior and Evolution," *Trends in Ecology & Evolution* 24, no. 8 (August 2009): 439–446, https://doi.org/10.1016/j.tree.2009.03.014.

29.   W. Byne and B. Parsons, "Human Sexual Orientation: The Biologic Theories Reappraised," *Archives of General Psychiatry* 50, no. 3 (March 1993): 228–239, https://doi.org/10.1001/archpsyc.1993.01820150078009.

14. PRONOUNS

1.   S. Stryker, *Transgender History*, 2nd ed. (New York: Seal Press, 2017).

2.   M. Goodman, N. Adams, T. Cornell, et al., "Size of Transgender and Gender-nonconforming Populations," *Endocrinology & Metabolism Clinics of North America* 48 (2019): 303–321, https://doi.org/10.1016/j.ecl.2019.01.001.

3.   Q. Zhang, M. Goodman, N. Adams, et al., "Epidemiological Considerations in Transgender Health: A Systematic Review with Focus on Higher Quality Data," *International Journal of Transgenderism* 21 (2020): 125–137, https://doi.org/10.1080/26895269.2020.1753136.

4.   P. Saisuwan, "Kathoey and the Linguistic Construction of Gender Identity in Thailand," in *Language, Sexuality, and Power: Studies in Intersectional Sociolinguistics, Studies in Language Gender and Sexuality*, ed. Erez Levon and Ronald Beline Mendes (New York: Oxford University Press, 2015), 189–214, https://doi.org/10.1093/acprof:oso/9780190210366.003.0010.

5.   G. Warnke, *Debating Sex and Gender* (New York: Oxford University Press, 2011).

6.   N. J. Bradford, G. N. Rider, and K. G. Spencer, "Hair Removal and Psychological Well-Being in Transfeminine Adults: Associations with Gender Dysphoria and Gender Euphoria," *Journal of Dermatological Treatment* (2019): 1–8, https://doi.org/10.1080/09546634.2019.1687823; N. J. Bradford and M. Syed, "Transnormativity and Transgender Identity Development: A Master Narrative Approach," *Sex Roles* 81 (2019): 306–325, https://doi.org/10.1007/s11199-018-0992-7.

7.   W. J. Beischel, S. E. M. Gauvin, and S. M. van Anders, "'A Little Shiny Gender Breakthrough': Community Understandings of Gender Euphoria," *International Journal of Transgender Health* 23, no. 3 (2022): 274–294, https://doi.org/10.1080/26895269.2021.1915223.

8.   S. E. James, J. L. Herman, S. Rankin, M. Keisling, L. Mottet, and M. Anafi, *The Report of the 2015 U.S. Transgender Survey* (Washington, DC: National Center for Transgender Equality, 2016).

9.   L. Littman, "Parent Reports of Adolescents and Young Adults Perceived to Show Signs of a Rapid Onset of Gender Dysphoria," *PLoS One* 13, no. 8 (2018): e0202330, https://doi.org/10.1371/journal.pone.0202330.

10.   A. Brandelli Costa, "Formal Comment On: Parent Reports of Adolescents and Young Adults Perceived to Show Signs of a Rapid Onset of Gender Dysphoria," *PLoS One* 14, no. 3 (2019): e0212578, https://doi.org/10.1371/journal.pone.0212578; A. J. Restar, "Methodological Critique of Littman's (2018) Parental-Respondents Accounts of 'Rapid-Onset Gender Dysphoria,'" *Archives of Sexual Behavior* 49, no. 1 (2019) 61–66, https://doi.org/10.1007/s10508-019-1453-2.

11.   Coalition for the Advancement & Application of Psychological Science, "CAAPS Position Statement on Rapid Onset Gender Dysphoria (ROGD)," 2001, https://www.caaps.co/rogd-statement.

12. American Psychiatric Association, *Diagnostic and Statistical Manual of Mental Disorders: DSM-5-Tr*, 5th ed., text rev. ed. (Washington, DC: American Psychiatric Association, 2022).

13. A. Kirzinger, A. Kearney, A. Montero, G. Sparks, L. Dawson, and M. Brodie, *KFF/ The Washington Post Trans Survey* (*Washington Post*/KFF Survey Project, March 2022), https://files.kff.org/attachment/REPORT-KFF-The-Washington-Post-Trans-Survey.pdf (accessed February 26, 2024).

14. S. Hattenstone, "The Dad Who Gave Birth: 'Being Pregnant Doesn't Change Me Being a Trans Man,'" *The Guardian*, April 20, 2019.

15. A. N. Almazan and A. S. Keuroghlian, "Association between Gender-Affirming Surgeries and Mental Health Outcomes," *JAMA Surgery* 156, no. 7 (July 1, 2021): 611–618, https://doi.org/10.1001/jamasurg.2021.0952.

16. S. Cachero, A. D. Ostrovsky, J. Y. Yu, B. J. Dickson, and G. S. Jefferis, "Sexual Dimorphism in the Fly Brain," *Current Biology* 20, no. 18 (September 28, 2010): 1589–1601, https://doi.org/10.1016/j.cub.2010.07.045; E. Demir and B. J. Dickson, "Fruitless Splicing Specifies Male Courtship Behavior in Drosophila," *Cell* 121, no. 5 (June 3, 2005): 785–794, https://doi.org/10.1016/j.cell.2005.04.027.

17. S. LeVay, "A Difference in Hypothalamic Structure between Heterosexual and Homosexual Men," *Science* 253, no. 5023 (August 30, 1991): 1034–1037, https://doi.org/10.1126/science.1887219.

18. W. Byne, M. S. Lasco, E. Kemether, A. Shinwari, M. A. Edgar, S. Morgello, et al., "The Interstitial Nuclei of the Human Anterior Hypothalamus: An Investigation of Sexual Variation in Volume and Cell Size, Number and Density," *Brain Research* 856 (2000): 254–258, https://doi.org/10.1006/hbeh.2001.1680.

19. A. Garcia-Falgueras and D. F. Swaab, "A Sex Difference in the Hypothalamic Uncinate Nucleus: Relationship to Gender Identity," *Brain* 131, pt. 12 (December 2008): 3132–3146, https://doi.org/10.1093/brain/awn276.

20. A. Guillamon, C. Junque, and E. Gómez-Gil, "A Review of the Status of Brain Structure Research in Transsexualism," *Archives of Sexual Behavior* 45, no. 7 (October 2016): 1615–1648, https://doi.org/10.1007/s10508-016-0768-5; B. Carrillo, E. Gómez-Gil, G. Rametti, C. Junque, Á. Gomez, K. Karadi, S. Segovia, and A. Guillamon, "Cortical Activation During Mental Rotation in Male-to-Female and Female-to-Male Transsexuals Under Hormonal Treatment," *Psychoneuroendocrinology* 35, no. 8 (September 2010): 1213–1222, https://doi.org/https://doi.org/10.1016/j.psyneuen.2010.02.010.

21. M. B. Reed, P. A. Handschuh, M. Klöbl, M. E. Konadu, U. Kaufmann, A. Hahn, G. S. Kranz, M. Spies, and R. Lanzenberger, "The Influence of Sex Steroid Treatment on Insular Connectivity in Gender Dysphoria," *Psychoneuroendocrinology* 155 (September 2023): 106336, https://doi.org/10.1016/j.psyneuen.2023.106336.

EPILOGUE: EYES WIDE OPEN

1. M. Mikkola, "Feminist Perspectives on Sex and Gender," in *The Stanford Encyclopedia of Philosophy* (Spring 2023 ed.), ed. Edward N. Zalta and Uri Nodelman (Stanford, CA: Stanford University, 2023), https://plato.stanford.edu/archives/spr2023/entries/feminism-gender/.

2.   O. Oyewumi, *The Invention of Women* (Minneapolis: University of Minnesota Press, 1997).

3.   W. Dan, "The Search for Non-binary Pronouns in Chinese," *MultiLingual*, July 26, 2021, https://multilingual.com/the-search-for-non-binary-pronouns-in-chinese/.

4.   "What We Learned About Sexual Desire from 10 Years of Pornhub User Data," *The Cut*, June 11, 2017, https://www.thecut.com/2017/06/pornhub-data-sexual-habits.html.

5.   L. D. Lindberg and I. Maddow-Zimet, "Consequences of Sex Education on Teen and Young Adult Sexual Behaviors and Outcomes," *Journal of Adolescent Health* 51 (2012): 332–338, https://doi.org/10.1016/j.jadohealth.2011.12.028.

6.   For example, Florida's HB1557: Florida House of Representatives, 2022 Legislature, https://www.flsenate.gov/Session/Bill/2022/1557/BillText/er/PDF.

7.   W. Lenz, "A Short History of Thalidomide Embryopathy," *Teratology* 38 (1988): 203–215, https://doi.org/10.1002/tera.1420380303.

8.   U.S. Food and Drug Administration, *General Considerations for the Clinical Evaluation of Drugs* (Rockville, MD: U.S. Food and Drug Administration, 1977).

9.   K. A. Liu and N. A. DiPietro Mager, "Women's Involvement in Clinical Trials: Historical Perspective and Future Implications," *Pharmacy Practice* 14 (January–March 2016): 708, https://doi.org/10.18549/PharmPract.2016.01.708.

10.   National Institutes of Health, Inclusion of Women and Minorities as Participants in Research Involving Human Subjects, last updated October 11, 2022, https://grants.nih.gov/policy/inclusion/women-and-minorities.htm.

11.   A. H. Maas and Y. E. Appelman, "Gender Differences in Coronary Heart Disease," *Netherlands Heart Journal* 18 (December 2010): 598–603, https://doi.org/10.1007/s12471-010-0841-y; S. A. E. Peters, P. Muntner, and M. Woodward, "Sex Differences in the Prevalence of, and Trends in, Cardiovascular Risk Factors, Treatment, and Control in the United States, 2001 to 2016," *Circulation* 139 (February 2019): 1025–1035, https://doi.org/10.1161/CIRCULATIONAHA.118.035550.

12.   H. Kleven, et al., "Children and Gender Inequality: Evidence from Denmark." National Bureau of Economic Research, Working Paper 24219, National Bureau of Economic Research, 2018, https://www.nber.org/papers/w24219.

13.   T. Valerio, et al., "Childless Older Americans: 2018," U.S. Census Bureau, *Current Population Reports* (Washington, DC: U.S. Government Publishing Office, 2021), 70–173, https://www.census.gov/content/dam/Census/library/publications/2021/demo/p70-173.pdf.

Bibliography

Abramson, R. G., A. Mavi, T. Cermik, S. Basu, N. E. Wehrli, M. Houseni, S. Mishra, J. Udupa, P. Lakhani, A. D. Maidment, D. A. Torigian, and A. Alavi. 2007. "Age-Related Structural and Functional Changes in the Breast: Multimodality Correlation with Digital Mammography, Computed Tomography, Magnetic Resonance Imaging, and Positron Emission Tomography." *Seminars in Nuclear Medicine* 37, no. 3: 146–153. https://doi.org/10.1053/j.semnuclmed.2007.01.003.

Ah-King, M., and S. Nylin. 2010. "Sex in an Evolutionary Perspective: Just Another Reaction Norm." *Evolutionary Biology* 37, no. 4: 234–246. https://doi.org/10.1007/s11692-010-9101-8.

Almazan, A. N., and A. S. Keuroghlian. 2021. "Association Between Gender-Affirming Surgeries and Mental Health Outcomes." *JAMA Surgery* 156, no. 7: 611–618. https://doi.org/10.1001/jamasurg.2021.0952.

American Psychiatric Association. 2022. *Diagnostic and Statistical Manual of Mental Disorders: DSM-5-TR*. 5th ed., text rev. Arlington, VA: American Psychiatric Association.

Askew, G. N. 2014. "The Elaborate Plumage in Peacocks Is Not Such a Drag." *Journal of Experimental Biology* 217 (pt 18): 3237–3241. https://doi.org/10.1242/jeb.107474.

Bachtrog, D., J. E. Mank, C. L. Peichel, M. Kirkpatrick, S. P. Otto, T. L. Ashman, M. W. Hahn, J. Kitano, I. Mayrose, R. Ming, N. Perrin, L. Ross, N. Valenzuela, J. C. Vamosi, and Consortium Tree of Sex. 2014. "Sex Determination: Why So Many Ways of Doing It?" *PLoS Biology* 12, no. 7: e1001899. https://doi.org/10.1371/journal.pbio.1001899.

Bailey, J. M., M. P. Dunne, and N. G. Martin. 2000. "Genetic and Environmental Influences on Sexual Orientation and Its Correlates in an Australian Twin Sample." *Journal of Personality and Social Psychology* 78, no. 3: 524–536. https://doi.org/10.1037//0022-3514.78.3.524.

Bailey, J. M., and K. J. Zucker. 1995. "Childhood Sex-Typed Behavior and Sexual Orientation: A Conceptual Analysis and Quantitative Review." *Developmental Psychology* 31, no. 1: 43–55. https://doi.org/10.1037/0012-1649.31.1.43.

Bailey, N. W., and M. Zuk. 2009. "Same-Sex Sexual Behavior and Evolution." *Trends in Ecology & Evolution* 24, no. 8: 439–446. https://doi.org/10.1016/j.tree.2009.03.014.

Bakker, J., and M. J. Baum. 2000. "Neuroendocrine Regulation of GnRH Release in Induced Ovulators." *Frontiers in Neuroendocrinology* 21, no. 3: 220–262. https://doi.org/10.1006/frne.2000.0198.

Barlow, D. P., and M. S. Bartolomei. 2014. "Genomic Imprinting in Mammals." *Cold Spring Harbor Perspectives in Biology* 6, no. 2: a018382 https://doi.org/10.1101/cshperspect.a018382.

Barr, M. L., and E. G. Bertram. 1949. "A Morphological Distinction between Neurones of the Male and Female, and the Behaviour of the Nucleolar Satellite during Accelerated Nucleoprotein Synthesis." *Nature* 163, no. 4148: 676. https://doi.org/10.1038/163676a0.

Baskin, L., J. Shen, A. Sinclair, M. Cao, X. Liu, G. Liu, D. Isaacson, M. Overland, Y. Li, and G. R. Cunha. 2018. "Development of the Human Penis and Clitoris." *Differentiation* 103: 74–85. https://doi.org/10.1016/j.diff.2018.08.001.

Beckerman, S., and P. Valentine, eds. 2002. *Cultures of Multiple Fathers: The Theory and Practice of Partible Paternity in Lowland South America*. Gainesville: University of Florida Press.

Beischel, W. J., S. E. M. Gauvin, and S. M. van Anders. 2022. " 'A Little Shiny Gender Breakthrough': Community Understandings of Gender Euphoria." *International Journal of Transgender Health* 23, no. 3: 274–294. https://doi.org/10.1080/26895269.2021.1915223.

Bell, G. 1978. "The Evolution of Anisogamy." *Journal of Theoretical Biology* 73, no. 2: 247–270. https://doi.org/10.1016/0022-5193(78)90189-3.

Berenbaum, S. A., and M. Hines. 1992. "Early Androgens Are Related to Childhood Sex-Typed Toy Preferences." *Psychological Science* 3, no. 3: 203–206. https://doi.org/10.1111/j.1467-9280.1992.tb00028.x.

Berglund, A., and G. Rosenqvist. 2001. "Male Pipefish Prefer Ornamented Females." *Animal Behaviour* 61, no. 2: 345–350. https://doi.org/https://doi.org/10.1006/anbe.2000.1599.

Berkane, N., P. Liere, J. P. Oudinet, A. Hertig, G. Lefevre, N. Pluchino, M. Schumacher, and N. Chabbert-Buffet. 2017. "From Pregnancy to Preeclampsia: A Key Role for Estrogens." *Endocrine Reviews* 38, no. 2: 123–144. https://doi.org/10.1210/er.2016-1065.

Betancourt, A. J., and D. C. Presgraves. 2002. "Linkage Limits the Power of Natural Selection in Drosophila." *Proceedings of the National Academy of Sciences of the United States of America* 99, no. 21: 13616–13620. https://doi.org/10.1073/pnas.212277199.

Bhatnagar, K. P., and E. Meisami. 1998. "Vomeronasal Organ in Bats and Primates: Extremes of Structural Variability and Its Phylogenetic Implications." *Microscopy Research and Technique* 43, no. 6: 465–375. https://doi.org/10.1002/(SICI)1097-0029(19981215)43:6<465::AID-JEMT1>3.0.CO;2-1.

Blanchard, R. 1997. "Birth Order and Sibling Sex Ratio in Homosexual Versus Heterosexual Males and Females." *Annual Review of Sex Research* 8: 27–67.

Blanchard, R. 2001. "Fraternal Birth Order and the Maternal Immune Hypothesis of Male Homosexuality." *Hormones and Behavior* 40, no. 2: 105–114. https://doi.org/10.1006/hbeh.2001.1681.

Blanchard, R., and A. F. Bogaert. 2004. "Proportion of Homosexual Men Who Owe Their Sexual Orientation to Fraternal Birth Order: An Estimate Based on Two National Probability Samples." *American Journal of Human Biology* 16, no. 2: 151–157. https://doi.org/10.1002/ajhb.20006.

Bloomfield, G. 2020. "The Molecular Foundations of Zygosis." *Cellular and Molecular Life Sciences* 77, no. 2: 323–330. https://doi.org/10.1007/s00018-019-03187-1.

Boschetti, C., A. Carr, A. Crisp, I. Eyres, Y. Wang-Koh, E. Lubzens, T. G. Barraclough, G. Micklem, and A. Tunnacliffe. 2012. "Biochemical Diversification through Foreign Gene Expression in Bdelloid Rotifers." *PLoS Genetics* 8, no. 11: e1003035. https://doi.org/10.1371/journal.pgen.1003035.

Bradford, N. J., G. Rider, and K. G. Spencer. 2021. "Hair Removal and Psychological Well-Being in Transfeminine Adults: Associations with Gender Dysphoria and Gender Euphoria." *Journal of Dermatological Treatment* 32, no. 6: 635–642. https://doi.org/10.1080/09546634.2019.1687823.

Bradford, Nova J., and Moin Syed. 2019. "Transnormativity and Transgender Identity Development: A Master Narrative Approach." *Sex Roles* 81, no. 5: 306–325. https://doi.org/10.1007/s11199-018-0992-7.

Brandelli C. A. 2019. "Formal Comment On: Parent Reports of Adolescents and Young Adults Perceived to Show Signs of a Rapid Onset of Gender Dysphoria." *PLoS One* 14, no. 3: e0212578. https://doi.org/10.1371/journal.pone.0212578.

Brawand, D., W. Wahli, and H. Kaessmann. 2008. "Loss of Egg Yolk Genes in Mammals and the Origin of Lactation and Placentation." *PLoS Biology* 6 (3): e63. https://doi.org/10.1371/journal.pbio.0060063.

Breed, W. G. 1996. "Egg Maturation and Fertilization in Marsupials." *Reproduction, Fertility and Development* 8, no. 4: 617–643. https://doi.org/10.1071/rd9960617.

Brennan, P. L. R., C. J. Clark, and R. O. Prum. 2010. "Explosive Eversion and Functional Morphology of the Duck Penis Supports Sexual Conflict in Waterfowl Genitalia." *Proceedings of the Royal Society of London, Series B: Biological Sciences* 277, no. 1686: 1309–1314. https://doi.org/10.1098/rspb.2009.2139.

Bridges, C. B. 1916. "Non-Disjunction as Proof of the Chromosome Theory of Heredity." *Genetics* 1, no. 1: 1–52. https://doi.org/10.1093/genetics/1.1.1.

Brownmiller, S. 1975. *Against Our Will: Men, Women, and Rape.* New York: Fawcett.

Brush, S. G. 1978. "Nettie M. Stevens and the Discovery of Sex Determination by Chromosomes." *Isis* 69, no. 2: 163–172.

Buckley, T., and A. Gottlieb, eds. 1988. *Blood Magic: The Anthropology of Menstruation* Berkeley: University of California Press.

U.S. Food and Drug Administration (U.S. FDA) (ed.). 1977. "General Considerations for the Clinical Evaluation of Drugs." U.S. Food and Drug Administration. Rockville, MD: U.S. FDA, Bureau of Drugs.

Byne, W., and B. Parsons. 1993. "Human Sexual Orientation: The Biologic Theories Reappraised." *Archives of General Psychiatry* 50, no. 3: 228–239. https://doi.org/10.1001/archpsyc.1993.01820150078009.

Byne, W., S. Tobet, L. A. Mattiace, M. S. Lasco, E. Kemether, M. A. Edgar, S. Morgello, M. S. Buchsbaum, and L. B. Jones. 2001. "The Interstitial Nuclei of the Human Anterior Hypothalamus: An Investigation of Variation with Sex, Sexual Orientation, and HIV Status." *Hormones and Behavior* 40, no. 2: 86–92. https://doi.org/10.1006/hbeh.2001.1680.

Cachero, S., A. D. Ostrovsky, J. Y. Yu, B. J. Dickson, and G. S. Jefferis. 2010. "Sexual Dimorphism in the Fly Brain." *Current Biology* 20, no. 18: 1589–1601. https://doi.org/10.1016/j.cub.2010.07.045.

Callomon, J. H. 1963. *Sexual Dimorphism in Jurassic Ammonites. Transactions of the Leicester Literary and Philosophical Society,* vol. 57. Leicester, UK: W. Thornley & Son, Leicester.

Camperio Ciani, A. S., L. Fontanesi, F. Iemmola, E. Giannella, C. Ferron, and L. Lombardi. 2012. "Factors Associated with Higher Fecundity in Female Maternal Relatives of Homosexual Men." *Journal of Sexual Medicine* 9, no. 11: 2878–2887. https://doi.org/10.1111/j.1743-6109.2012.02785.x.

Carrillo, B., E. Gómez-Gil, G. Rametti, C. Junque, Á. Gomez, K. Karadi, S. Segovia, and A Guillamon. 2010. "Cortical Activation during Mental Rotation in Male-to-Female and Female-to-Male Transsexuals under Hormonal Treatment." *Psychoneuroendocrinology* 35, no. 8: 1213–1222. https://doi.org/https://doi.org/10.1016/j.psyneuen.2010.02.010.

Carroll, S. B. 2005. *Endless Forms Most Beautiful.* New York: Norton.

Castle, W. E. 1922. "The Y-Chromosome Type of Sex-Linked Inheritance in Man." *Science* 55 (1435): 703–704.

Cattanach, B. M., C. E. Pollard, and S. G. Hawker. 1971. "Sex-Reversed Mice: XX and XO Males." *Cytogenetics* 10, no. 5: 318–337. https://doi.org/10.1159/000130151.

Chamie, J. 2018. "Increasingly Indispensable Grandparents." YaleGlobal Online (September 4, 2018). https://archive-yaleglobal.yale.edu/content/increasingly-indispensable-grandparents.

Chan, R., D. Stuart-Fox, and T. S. Jessop. 2009. "Why Are Females Ornamented? A Test of the Courtship Stimulation and Courtship Rejection Hypotheses." *Behavioral Ecology* 20, no. 6: 1334–1342. https://doi.org/10.1093/beheco/arp136.

Chandra, A., W. D. Mosher, C. Copen, and C. Sionean. 2011. "Sexual Behavior, Sexual Attraction, and Sexual Identity in the United States: Data from the 2006–2008 National Survey of Family Growth." *National Health Statistics Reports* no. 36: 1–36.

Chapais, B. 2008. *Primeval Kinship: How Pair-Bonding Gave Birth to Human Society* Cambridge, MA: Harvard University Press.

Charnov, E. L. 1982. *The Theory of Sex Allocation.* Ed. Robert M. May. Vol. 18.*Monographs in Population Biology.* Princeton, NJ: Princeton University Press.

Chavan, A. R., O. W. Griffith, and G. P. Wagner. 2017. "The Inflammation Paradox in the Evolution of Mammalian Pregnancy: Turning a Foe into a Friend." *Current Opinion in Genetics & Development* 47: 24–32. https://doi.org/10.1016/j.gde.2017.08.004.

Childs, G. 2001. "Old-Age Security, Religious Celibacy, and Aggregate Fertility in a Tibetan Population." *Journal of Population Research* 18, no. 1: 52–67. https://doi.org/10.1007/BF03031955.

Churchill Downs. 2021. "The History of Women Jockeys in the Kentucky Derby." *DerbyExperiences.com* (blog). February 5, 2021. https://derbyexperiences.com/blog/women-jockeys-in-the-kentucky-derby.

Cindrova-Davies, T., E. Jauniaux, M. G. Elliot, S. Gong, G. J. Burton, and D. S. Charnock-Jones. 2017. "RNA-Seq Reveals Conservation of Function among the Yolk Sacs of Human, Mouse, and Chicken." *Proceedings of the National Academy of Sciences of the United States of America* 114, no. 24: E4753–E4761. https://doi.org/10.1073/pnas.1702560114.

Clark, R. V., J. A. Wald, R. S. Swerdloff, C. Wang, F. C. W. Wu, L. D. Bowers, and A. M. Matsumoto. 2019. "Large Divergence in Testosterone Concentrations between Men and Women: Frame of Reference for Elite Athletes in Sex-Specific Competition in Sports, a Narrative Review." *Clinical Endocrinology (Oxf)* 90, no. 1: 15–22. https://doi.org/10.1111/cen.13840.

Cline, T. W. 1979. "A Male-Specific Lethal Mutation in *Drosophila melanogaster* That Transforms Sex." *Developmental Biology* 72, no. 2: 266–275. https://doi.org/10.1016/0012-1606(79)90117-9.

Coalition for the Advancement & Application of Psychological Science (CAAPS). 2021. "CAAPS Position Statement on Rapid Onset Gender Dysphoria (ROGD)." https://www.caaps.co/rogd-statement.

Cornelis, G., M. Funk, C. Vernochet, F. Leal, O. A. Tarazona, G. Meurice, O. Heidmann, A. Dupressoir, A. Miralles, M. P. Ramirez-Pinilla, and T. Heidmann. 2017. "An Endogenous

Retroviral Envelope Syncytin and Its Cognate Receptor Identified in the Viviparous Placental Mabuya Lizard." *Proceedings of the National Academy of Sciences of the United States of America* 114, no. 51: E10991–E11000. https://doi.org/10.1073/pnas.1714590114.

Cranford, T. W., M. Amundin, and K. S. Norris. 1996. "Functional Morphology and Homology in the Odontocete Nasal Complex: Implications for Sound Generation." *Journal of Morphology* 228, no. 3: 223–285. https://doi.org/10.1002/(SICI)1097-4687(199606)228:3<223::AID-JMOR1>3.0.CO;2-3.

Cunha, G. R., S. J. Robboy, T. Kurita, D. Isaacson, J. Shen, M. Cao, and L. S. Baskin. 2018. "Development of the Human Female Reproductive Tract." *Differentiation* 103: 46–65. https://doi.org/https://doi.org/10.1016/j.diff.2018.09.001.

D'Alba, L., D. N. Jones, H. T. Badawy, C. M. Eliason, and M. D. Shawkey. 2014. "Antimicrobial Properties of a Nanostructured Eggshell from a Compost-Nesting Bird." *Journal of Experimental Biology* 217, pt. 7: 1116–1121. https://doi.org/10.1242/jeb.098343.

D'Alba, L., R. Maia, M. E. Hauber, and M. D. Shawkey. 2016. "The Evolution of Eggshell Cuticle in Relation to Nesting Ecology." *Proceedings of the Royal Society of London, Series B: Biological Sciences* 283, no. 1836: 20160687. https://doi.org/10.1098/rspb.2016.0687.

Dan, W. 2021. The Search for Non-Binary Pronouns in Chinese. *MultiLingual*. July 26, 2021. https://multilingual.com/the-search-for-non-binary-pronouns-in-chinese/.

Darroch, S. A. F., E. F. Smith, M. Laflamme, and D. H. Erwin. 2018. "Ediacaran Extinction and Cambrian Explosion." *Trends in Ecology & Evolution* 33, no. 9: 653–663. https://doi.org/10.1016/j.tree.2018.06.003.

Darwin, C. 1854. *A Monograph On the Sub-Class Cirripedia (The Balanidae and The Verrucidae)*. London: Ray Society.

Darwin, C. 1859. *On the Origin of Species by Means of Natural Selection*. London: John Murray.

Deady, D. K., M. J. Law Smith, J. P. Kent, and R. I. M. Dunbar. 2006. "Is Priesthood an Adaptive Strategy? Evidence from a Historical Iriship Population." *Human Nature* 17, no. 4: 393–404. https://doi.org/https://doi.org/10.1007/s12110-006-1002-2.

Deines, T. 2018. "Why Snakes Have Two Penises and Alligators Are Always Erect." *National Geographic News*: https://www.nationalgeographic.com/news/2018/02/snakes-alligators-reptiles-genitalia-animals/.

Delachapelle, A., H. Hortling, M. Niemi, and J. Wennstroem. 1964. "XX Sex Chromosomes in a Human Male: First Case." *Acta Medica Scandinavica* 175 (Suppl. 412): 25–28. https://doi.org/10.1111/j.0954-6820.1964.tb04630.x.

Delwiche, C. F., and R. E. Timme. 2011. "Plants." *Current Biology* 21, no. 11: R417–R422. https://doi.org/10.1016/j.cub.2011.04.021.

Demir, E., and B. J. Dickson. 2005. "Fruitless Splicing Specifies Male Courtship Behavior in Drosophila." *Cell* 121, no. 5: 785–794. https://doi.org/10.1016/j.cell.2005.04.027.

Diamond, Lisa. 2023. "What Develops in the Biodevelopment of Sexual Orientation?" *Archives of Sexual Behavior* 52. https://doi.org/10.1007/s10508-023-02542-5.

Diaz, O. 2023. "Rise in Strangulations Worries Fairfax County." *Washington Post*, March 7, 2023, Metro.

Dixon, A. F. 2009. "Human Sexual Dimorphism: Opposites Attract," Chapter 7 in *Sexual Selection and the Origins of Human Mating Systems*, Oxford: Oxford University Press.

Dixson, A. 1998. *Primate Sexuality: Comparative Studies of the Prosimians, Monkeys, Apes and Human Beings*. Oxford: Oxford Univesity Press.

Doerr, P., G. Kockott, H. J. Vogt, K. M. Pirke, and F. Dittmar. 1973. "Plasma Testosterone, Estradiol, and Semen Analysis in Male Homosexuals." *Archives of General Psychiatry* 29, no. 6: 829–833. https://doi.org/10.1001/archpsyc.1973.04200060101016.

Dolgin, E. S., B. Charlesworth, S. E. Baird, and A. D. Cutter. 2007. "Inbreeding and Outbreeding Depression in Caenorhabditis Nematodes." *Evolution* 61, no. 6: 1339–1352.

Dunlap, K. A., M. Palmarini, M. Varela, R. C. Burghardt, K. Hayashi, J. L. Farmer, and T. E. Spencer. 2006. "Endogenous Retroviruses Regulate Periimplantation Placental Growth and Differentiation." *Proceedings of the National Academy of Sciences of the United States of America* 103, no. 39: 14390–14395. https://doi.org/10.1073/pnas.0603836103.

Dunsworth, H., and L. Eccleston. 2015. "The Evolution of Difficult Childbirth and Helpless Hominin Infants." *Annual Review of Anthropology* 44, no. 1: 55–69. https://doi.org/10.1146/annurev-anthro-102214-013918.

Dupressoir, A., G. Marceau, C. Vernochet, L. Benit, C. Kanellopoulos, V. Sapin, and T. Heidmann. 2005. "Syncytin-A and Syncytin-B, Two Fusogenic Placenta-Specific Murine Envelope Genes of Retroviral Origin Conserved in Muridae." *Proceedings of the National Academy of Sciences of the United States of America* 102, no. 3: 725–730. https://doi.org/10.1073/pnas.0406509102.

Eberhard, W. 1985. *Sexual Selection and Animal Genitalia*. Princeton, NJ: Princeton University Press.

Edwards, D., and P. Kenrick. 2015. "The Early Evolution of Land Plants, from Fossils to Genomics: A Commentary on Lang (1937) 'On the Plant-Remains from the Downtonian of England and Wales.'" *Philosophical Transactions of the Royal Society of London, Series B: Biological Sciences* 370, no. 1666. https://doi.org/10.1098/rstb.2014.0343.

Eldridge, N. 2004. *Why We Do It*. New York: Norton.

Ellibeş Kaya, A., O. Doğan, M. Yassa, A. Başbuğ, C. Özcan, and E. Çalışkan. 2020. "Do External Female Genital Measurements Affect Genital Perception and Sexual Function and Orgasm?" *Turkish Journal of Obstetrics and Gynecology* 17, no. 3: 175–181. https://doi.org/10.4274/tjod.galenos.2020.89896.

Emera, D., R. Romero, and G. Wagner. 2012. "The Evolution of Menstruation: A New Model for Genetic Assimilation." *Bioessays* 34, no. 1: 26–35. https://doi.org/10.1002/bies.201100099.

Emlen, S. T., and P. H. Wrege. 2004. "Size Dimorphism, Intrasexual Competition, and Sexual Selection in Wattled Jacana (*Jacana jacana*), A Sex-Role-Reversed Shorebird in Panama." *The Auk* 121, no. 2: 391–403. https://doi.org/10.1093/auk/121.2.391.

Espada, M. 2018. "Period Taboo Around the World." *Teen Vogue*. https://www.teenvogue.com/story/period-taboo-around-the-world. Accessed February, 2024.

Fales, Melissa R., David A. Frederick, Justin R. Garcia, Kelly A. Gildersleeve, Martie G. Haselton, and Helen E. Fisher. 2016. "Mating Markets and Bargaining Hands: Mate Preferences for Attractiveness and Resources in Two National U.S. Studies." *Personality and Individual Differences* 88: 78–87. https://doi.org/https://doi.org/10.1016/j.paid.2015.08.041.

Fang, H., C. M. Disteche, and J. B. Berletch. 2019. "X Inactivation and Escape: Epigenetic and Structural Features." *Frontiers in Cell Development and Biology* 7: 219. https://doi.org/10.3389/fcell.2019.00219.

Fausto-Sterling, A. 2018. "Why Sex Is Not Binary." *New York Times*, October 29, 2018.

Fedry, J., Y. Liu, G. Pehau-Arnaudet, J. Pei, W. Li, M. A. Tortorici, F. Traincard, A. Meola, G. Bricogne, N. V. Grishin, W. J. Snell, F. A. Rey, and T. Krey. 2017. "The Ancient Gamete

Fusogen HAP2 Is a Eukaryotic Class II Fusion Protein." *Cell* 168, no. 5: 904–915. https://doi.org/10.1016/j.cell.2017.01.024.

Fine, C. 2010. *Delusions of Gender*. New York: Norton.

Fischer, E. A. 1980. "The Relationship between Mating System and Simultaneous Hermaphroditism in the Coral Reef Fish, *Hypoplectrus nigricans* (Serranidae)." *Animal Behaviour* 28, no. 2: 620–633. https://doi.org/https://doi.org/10.1016/S0003-3472(80)80070-4.

Fisher, M. L., and C. Candea. 2012. "You Ain't Woman Enough to Take My Man: Female Intrasexual Competition as Portrayed in Songs." *Journal of Social, Evolutionary, and Cultural Psychology* 6, no. 4: 480–493. https://doi.org/10.1037/h0099238.

Fisher, M. L., and A. Cox. 2011. "Four Strategies Used during Intrasexual Competition for Mates." *Personal Relationships* 18, no. 1: 20–38. https://doi.org/https://doi.org/10.1111/j.1475-6811.2010.01307.x.

Flot, J. F., B. Hespeels, X. Li, B. Noel, I. Arkhipova, E. G. Danchin, A. Hejnol, B. Henrissat, R. Koszul, J. M. Aury, V. Barbe, R. M. Barthelemy, J. Bast, G. A. Bazykin, O. Chabrol, A. Couloux, M. Da Rocha, C. Da Silva, E. Gladyshev, P. Gouret, O. Hallatschek, B. Hecox-Lea, K. Labadie, B. Lejeune, O. Piskurek, J. Poulain, F. Rodriguez, J. F. Ryan, O. A. Vakhrusheva, E. Wajnberg, B. Wirth, I. Yushenova, M. Kellis, A. S. Kondrashov, D. B. Mark Welch, P. Pontarotti, J. Weissenbach, P. Wincker, O. Jaillon, and K. Van Doninck. 2013. "Genomic Evidence for Ameiotic Evolution in the Bdelloid Rotifer *Adineta vaga*." *Nature* 500, no 7463: 453–457. https://doi.org/10.1038/nature12326.

Ford, C. E., K. W. Jones, P. E. Polani, J. C. De Almeida, and J. H. Briggs. 1959. "A Sex-Chromosome Anomaly in a Case of Gonadal Dysgenesis (Turner's Syndrome)." *Lancet* 1, no. 7075: 711–713. https://doi.org/10.1016/s0140-6736(59)91893-8.

Frayer, D. W., and M. H. Wolpoff. 1985. "Sexual Dimorphism." *Annual Review of Anthropology* 14, no. 1: 429–473. https://doi.org/10.1146/annurev.an.14.100185.002241.

Fruth, B., and G. Hohmann. 2006. "Social Grease for Females? Same-Sex Genital Contacts in Wild Bonobos." In *Homosexual Behaviour in Animals: An Evolutionary Perspective*, ed. V. Sommer and P. L. Vasey, 294–315. New York: Cambridge University Press.

Fu, W., R. M. Gittelman, M. J. Bamshad, and J. M. Akey. 2014. "Characteristics of Neutral and Deleterious Protein-Coding Variation among Individuals and Populations." *American Journal of Human Genetics* 95, no. 4: 421–436. https://doi.org/10.1016/j.ajhg.2014.09.006.

Furman, E. R., and A. B. Yule. 1990. "Self-Fertilisation in *Balanus improvisus* Darwin." *Journal of Experimental Marine Biology and Ecology* 144, no. 2: 235–239. https://doi.org/https://doi.org/10.1016/0022-0981(90)90030-G.

Ganetzky, B., and R. S. Hawley. 2016. "The Centenary of GENETICS: Bridges to the Future." *Genetics* 202, no. 1: 15–23. https://doi.org/10.1534/genetics.115.180182.

Ganna, A., K. J. H. Verweij, M. G. Nivard, R. Maier, R. Wedow, A. S. Busch, A. Abdellaoui, S. Guo, J. F. Sathirapongsasuti, Team andMe Research, P. Lichtenstein, S. Lundstrom, N. Langstrom, A. Auton, K. M. Harris, G. W. Beecham, E. R. Martin, A. R. Sanders, J. R. B. Perry, B. M. Neale, and B. P. Zietsch. 2019. "Large-Scale GWAS Reveals Insights into the Genetic Architecture of Same-Sex Sexual Behavior." *Science* 365, no. 6456. https://doi.org/10.1126/science.aat7693.

Garcia, L. R., B. LeBoeuf, and P. Koo. 2007. "Diversity in Mating Behavior of Hermaphroditic and Male-Female *Caenorhabditis* Nematodes." *Genetics* 175, no. 4: 1761–1771. https://doi.org/10.1534/genetics.106.068304.

Garcia-Falgueras, A., and D. F. Swaab. 2008. "A Sex Difference in the Hypothalamic Unci-nate Nucleus: Relationship to Gender Identity." *Brain* 131, no. 12: 3132–3146. https://doi.org/10.1093/brain/awn276.

Geitler, L. 1932. *Der Formwechsel der pennaten Diatomeen (Kieselalgen)*. Jena: G. Fischer.

Georgetown Institute for Women, Peace and Security. 2022. "Mahram: Women's Mobility in Islam." *Voices for Change and Inclusion*. https://giwps.georgetown.edu/wp-content/uploads/2022/08/Mahram-Womens-Mobility-in-Islam.pdf.

Gershoni, M., and S. Pietrokovski. 2017. "The Landscape of Sex-Differential Transcriptome and Its Consequent Selection in Human Adults." *BMC Biology* 15, no. 1: 7. https://doi.org/10.1186/s12915-017-0352-z.

Gibbens, S. 2018. "Growing Population of Crayfish Has One Female Ancestor." *National Geographic Online*: https://www.nationalgeographic.com/animals/article/marbled-crayfish-marmorkrebs-evolution-genes-tumors-spd. Accessed February, 2024.

Gilbert, S.F. 2000. *Developmental Biology*. 6th ed. Sunderland, MA: Sinauer.

Gladyshev, E. A., M. Meselson, and I. R. Arkhipova. 2008. "Massive Horizontal Gene Transfer in Bdelloid Rotifers." *Science* 320, no. 5880: 1210–1213. https://doi.org/10.1126/science.1156407.

Goldberg, E. E., J. R. Kohn, R. Lande, K. A. Robertson, S. A. Smith, and B. Igic. 2010. "Species Selection Maintains Self-Incompatibility." *Science* 330, no. 6003: 493–495.

Goodenough, U., H. Lin, and J. H. Lee. 2007. "Sex Determination in Chlamydomonas." *Seminars in Cell and Developmental Biology* 18, no. 3: 350–361. https://doi.org/10.1016/j.semcdb.2007.02.006.

Goodman, M., N. Adams, T. Corneil, B. Kreukels, J. Motmans, and E. Coleman. 2019. "Size and Distribution of Transgender and Gender Nonconforming Populations: A Narrative Review." *Endocrinology and Metabolism Clinics of North America* 48, no. 2: 303–321. https://doi.org/10.1016/j.ecl.2019.01.001.

Gould, S. J. 1970. "Dollo on Dollo's Law: Irreversibility and the Status of Evolutionary Laws." *Journal of the History of Biology* 3: 189–212. https://doi.org/10.1007/BF00137351.

Gowaty, P. 2018. "Rape, Forced and Aggressively Coerced Copulation." In *Encyclopedia of Animal Behavior*, ed. J. C. Choe, 447–452. Cambridge, MA: Academic Press.

Graham-Harrison, E. 2022. " 'I Daren't Go Far': Taliban Rules Trap Afghan Women with No Male Guardian." *The Guardian*, August 15, 2022.

Gredler, M. L., C. E. Larkins, F. Leal, A. K. Lewis, A. M. Herrera, C. L. Perriton, T. J. Sanger, and M. J. Cohn. 2014. "Evolution of External Genitalia: Insights from Reptilian Development." *Sexual Development* 8, no. 5: 311–326. https://doi.org/10.1159/000365771.

Green, R. 1987. *The "Sissy Boy Syndrome" and the Development of Human Homosexuality*. New Haven, CT: Yale University Press.

Griffith, O. W., A. R. Chavan, S. Protopapas, J. Maziarz, R. Romero, and G. P. Wagner. 2017. "Embryo Implantation Evolved from an Ancestral Inflammatory Attachment Reaction." *Proceedings of the National Academy of Sciences of the United States of America* 114, no. 32: E6566–E6575. https://doi.org/10.1073/pnas.1701129114.

Grossnickle, D. M., S. M. Smith, and G. P. Wilson. 2019. "Untangling the Multiple Ecological Radiations of Early Mammals." *Trends in Ecology & Evolution* 34, no. 10: 936–949. https://doi.org/10.1016/j.tree.2019.05.008.

Grueter, C. C., and T. S. Stoinski. 2016. "Homosexual Behavior in Female Mountain Goril-las: Reflection of Dominance, Affiliation, Reconciliation or Arousal?" *PLoS One* 11, no. 5: e0154185. https://doi.org/10.1371/journal.pone.0154185.

Grutzner, F., W. Rens, E. Tsend-Ayush, N. El-Mogharbel, P. C. O'Brien, R. C. Jones, M. A. Ferguson-Smith, and J. A. Marshall Graves. 2004. "In the Platypus a Meiotic Chain of Ten Sex Chromosomes Shares Genes with the Bird Z and Mammal X Chromosomes." *Nature* 432, no. 7019: 913–917. https://doi.org/10.1038/nature03021.

Guay, A., and S. R. Davis. 2002. "Testosterone Insufficiency in Women: Fact or Fiction?" *World Journal of Urology* 20, no. 2: 106–110. https://doi.org/10.1007/s00345-002-0267-2.

Guillamon, A., C. Junque, and E. Gómez-Gil. 2016. "A Review of the Status of Brain Structure Research in Transsexualism." *Archives of Sexual Behavior* 45, no. 7: 1615–1648. https://doi.org/10.1007/s10508-016-0768-5.

Gutekunst, J., R. Andriantsoa, C. Falckenhayn, K. Hanna, W. Stein, J. Rasamy, and F. Lyko. 2018. "Clonal Genome Evolution and Rapid Invasive Spread of the Marbled Crayfish." *Nature Ecology & Evolution* 2, no. 3: 567–573. https://doi.org/10.1038/s41559-018-0467-9.

Guttes, E., and S. Guttes. 1964. "Thymidine Incorporation by Mitochondria in Physarum Polycephalum." *Science* 145, no. 3636: 1057–1058. https://doi.org/10.1126/science.145.3636.1057.

Haag, E. S. 2007. "Why Two Sexes? Sex Determination in Multicellular Organisms and Protistan Mating Types." *Seminars in Cell and Developmental Biology* 18, no. 3: 348–349.

Haag, E. S., J. Helder, P. J. Mooijman, D. Yin, and S. Hu. 2019. "The Evolution of Uniparental Reproduction in Rhabditina Nematodes: Phylogenetic Patterns, Developmental Causes, and Surprising Consequences." In *Transitions Between Sexual Systems*, ed. J. L. Leonard, 99–122. Cham, Switzerland: Spring Nature Switzerland AG.

Haag, E. S., and T. W. Lo. 2018. "How to Make a Billion Parasites." *Developmental Cell* 45 (2): 147–148. https://doi.org/10.1016/j.devcel.2018.04.006.

Hagstrom, E., and S. G. Andersson. 2018. "The Challenges of Integrating Two Genomes in One Cell." *Current Opinion in Microbiology* 41: 89–94. https://doi.org/10.1016/j.mib.2017.12.003.

Hakim, C. 2011. *Erotic Capital*. New York: Basic Books.

Hale, G., T. Regev, and Y. Rubinstein. 2021. "Do Looks Matter for an Academic Career in Economics?" CEPR Discussion Papers, no. 15893. Center for Economic Policy Research, London. https://ideas.repec.org/p/cpr/ceprdp/15893.html.

Hannema, S. E., and I. A. Hughes. 2007. "Regulation of Wolffian Duct Development." *Hormone Research* 67, no. 3: 142–151. https://doi.org/10.1159/000096644.

Hannibal, J. T., N. A. Reser, J. A. Yeakley, T. A. Kalka, and V. Fusco. 2013. "Determining Provenance of Local and Imported Chert Millstones Using Fossils (Especially Charophyta, Fusulinina, and Brachiopoda): Examples from Ohio, U.S.A.," *PALAIOS* 28, no. 11: 739–754, https://doi.org/10.2110/palo.2013.110.

Harrington, R. W., Jr. 1961. "Oviparous Hermaphroditic Fish with Internal Self-Fertilization." *Science* 134, no. 3492: 1749–1750. https://doi.org/10.1126/science.134.3492.1749.

Harrington, R. W. 1967. "Environmentally Controlled Induction of Primary Males Gonochorists from Eggs of the Self-Fertilizaing Hermaphroditic Fish, *Rivulus marmoratus* Poey." *Biological Bulletin*. 131: 174–199.

Harrington, R. W. 1971. "How Ecological and Genetic Factors Interact to Determine When Self-Fertilizing Hermaphrodites of *Rivulus marmoratus* Change into Functional Secondary Males, with a Reappraisal of the Modes of Intersexuality among Fishes." *Copeia* 1971, no. 3: 389–432.

Hattenstone, S. 2019. "The Dad Who Gave Birth: 'Being Pregnant Doesn't Change Me Being a Trans Man.'" *The Guardian*, April 20, 2019.

Hawaii Invasive Species Council. 2024. "Coquí." State of Hawaii. https://dlnr.hawaii.gov/hisc /info/invasive-species-profiles/coqui/. Accessed February 15, 2024.

Hayward, A., and J. F. Gillooly. 2011. "The Cost of Sex: Quantifying Energetic Investment in Gamete Production by Males and Females." *PLoS One* 6, no. 1: e16557. https://doi.org/10.1371 /journal.pone.0016557.

Heinsohn, R. 2008. "Ecology and Evolution of the Enigmatic Eclectus Parrot (*Eclectus roratus*)." *Journal of Avian Medicine and Surgery* 22, no. 2: 146–150. https://doi.org/10.1647/2007-031.1.

Heinsohn, R., S. Legge, and J. A. Endler. 2005. "Extreme Reversed Sexual Dichromatism in a Bird without Sex Role Reversal." *Science* 309, no. 5734: 617–619. https://doi.org/10.1126 /science.1112774.

Hejnol, A., M. Obst, A. Stamatakis, M. Ott, G. W. Rouse, G. D. Edgecombe, P. Martinez, et al. 2009. "Assessing the Root of Bilaterian Animals with Scalable Phylogenomic Methods." *Proceedings of the Royal Society B* 276, no. 1677: 4261–4270, https://doi.org/10.1098/rspb.2009 .0896.

Herrera, A. M., S. G. Shuster, C. L. Perriton, and M. J. Cohn. 2013. "Developmental Basis of Phallus Reduction during Bird Evolution." *Current Biology* 23, no. 12: 1065–1074. https://doi .org/10.1016/j.cub.2013.04.062.

Hill, W. G., and A. Robertson. 1966. "The Effect of Linkage on Limits to Artificial Selection." *Genetics Research* 8, no. 3: 269–294.

Hodges, W. L. 2004. "Evolution of Viviparity in Horned Lizards (Phrynosoma): Testing the Cold-Climate Hypothesis." *Journal of Evolutionary Biology* 17, no. 6: 1230–1237. https://doi .org/10.1111/j.1420-9101.2004.00770.x.

Hodgkin, J. 2002. "One Lucky XX Male: Isolation of the First *Caenorhabditis elegans* Sex-Determination Mutants." *Genetics* 162, no. 4: 1501–1504.

Hoover, S. E., C. I. Keeling, M. L. Winston, and K. N. Slessor. 2003. "The Effect of Queen Pheromones on Worker Honey Bee Ovary Development." *Naturwissenschaften* 90, no. 10: 477–480. https://doi.org/10.1007/s00114-003-0462-z.

Hrdy, S. B. 2013. "The 'One Animal in All Creation about Which Man Knows the Least.' " *Philosophical Transactions of the Royal Society of London, Series B: Biological Sciences* 368, no. 1631: 20130072. https://doi.org/10.1098/rstb.2013.0072.

Hughes, R. L., and L. S. Hall. 1998. "Early Development and Embryology of the Platypus." *Philosophical Transactions of the Royal Society of London, Series B: Biological Sciences* 353, no. 1372: 1101–1114. https://doi.org/10.1098/rstb.1998.0269.

Igic, B., L. Bohs, and J. R. Kohn. 2006. "Ancient Polymorphism Reveals Unidirectional Breeding System Shifts." *Proceedings of the National Academy of Sciences of the United States of America* 103, no. 5: 1359–1363. https://doi.org/10.1073/pnas.0506283103.

Imachi, H., M. K. Nobu, N. Nakahara, Y. Morono, M. Ogawara, Y. Takaki, Y. Takano, K. Uematsu, T. Ikuta, M. Ito, Y. Matsui, M. Miyazaki, K. Murata, Y. Saito, S. Sakai, C. Song, E. Tasumi, Y. Yamanaka, T. Yamaguchi, Y. Kamagata, H. Tamaki, and K. Takai. 2020. "Isolation of an Archaeon at the Prokaryote-Eukaryote Interface." *Nature* 577, no. 7791: 519–525. https://doi .org/10.1038/s41586-019-1916-6.

Ishikawa, F., and T. Naito. 1999. "Why Do We Have Linear Chromosomes? A Matter of Adam and Eve." *Mutation Research* 434, no. 2: 99–107. https://doi.org/10.1016/s0921-8777(99)00017-8.

Iwabe, N., K. Kuma, M. Hasegawa, S. Osawa, and T. Miyata. 1989. "Evolutionary Relationship of aAchaebacteria, Eubacteria, and Eukaryotes Inferred from Phylogenetic Trees of Duplicated

Genes." *Proceedings of the National Academy of Sciences of the United States of America* 86, no. 23: 9355–9359. https://doi.org/10.1073/pnas.86.23.9355.

Jablonski, N. G., and G. Chaplin. 2010. "Colloquium Paper: Human Skin Pigmentation as an Adaptation to UV Radiation." *Proceedings of the National Academy of Sciences of the United States of America* 107, Suppl 2: 8962–8968. https://doi.org/10.1073/pnas.0914628107.

Jacobs, P. A., and J. A. Strong. 1959. "A Case of Human Intersexuality Having a Possible XXY Sex-Determining Mechanism." *Nature* 183, no. 4657: 302–303. https://doi.org/10.1038/183302a0.

James, S. E., J. L. Herman, S. Rankin, M. Keisling, L. Mottet, and M. Anafi. 2016. *The Report of the 2015 U.S. Transgender Survey.* Washington, DC: National Center for Transgender Equality.

Janicke, T., I. K. Haderer, M. J. Lajeunesse, and N. Anthes. 2016. "Darwinian Sex Roles Confirmed across the Animal Kingdom." *Science Advances* 2, no. 2: e1500983. https://doi.org/10.1126/sciadv.1500983.

Janssen, I., S. B. Heymsfield, Z. M. Wang, and R. Ross. 2000. "Skeletal Muscle Mass and Distribution in 468 Men and Women Aged 18–88 Yr." *Journal of Applied Physiology* 89, no. 1: 81–88. https://doi.org/10.1152/jappl.2000.89.1.81.

Jarne, P., and J. R. Auld. 2006. "Animals Mix it Up Too: The Distribution of Self-Fertilization among Hermaphroditic Animals." *Evolution* 60, no. 9: 1816–1824.

Joel, D, and L. Vikhanski. 2019. *Gender Mosaic.* New York: Little, Brown Spark.

Kallee, E., and S. Okada. 1956. "Effect of X-Rays on Mitochondrial Desoxyribonuclease II." *Experimental Cell Research* 11, no. 1: 212–214. https://doi.org/10.1016/0014-4827(56)90206-3.

Kallman, K. D., and R. W. Harrington, Jr. 1964. "Evidence for the existence of homozygous clones in the self-fertilizing hermaphroditic Teleost *Rivulus marmoratus* (Poey)." *Biological Bulletin.* 126: 101–114.

Katju, V., E. M. LaBeau, K. J. Lipinski, and U. Bergthorsson. 2008. "Sex Change by Gene Conversion in a *Caenorhabditis elegans* fog-2 Mutant." *Genetics* 180, no. 1: 669–672.

Kenney, D. E., and G. G. Borisy. 2009. "Thomas Hunt Morgan at the Marine Biological Laboratory: Naturalist and Experimentalist." *Genetics* 181, no. 3: 841–846. https://doi.org/10.1534/genetics.109.101659.

Kerr, A. M., A. H. Baird, and T. P. Hughes. 2011. "Correlated Evolution of Sex and Reproductive Mode in Corals (Anthozoa: Scleractinia)." *Proceedings of the Royal Society of London, Series B: Biological Sciences* 278, no. 1702: 75–81. https://doi.org/10.1098/rspb.2010.1196.

Kielan-Jaworowska, Z. 1979. "Pelvic Structure and nature of Reproduction in Multituberculata." *Nature* 277, no. 5695: 402–403. https://doi.org/10.1038/277402a0.

Kimchi, T., J. Xu, and C. Dulac. 2007. "A Functional Circuit Underlying Male Sexual Behaviour in the Female Mouse Brain." *Nature* 448, no. 7157: 1009–1014. https://doi.org/10.1038/nature06089.

Kinsey, A. C., W. B. Pomeroy, and C. E. Martin. 1948. *Sexual Behavior in the Human Male.* Philadelphia: Saunders.

Kinsey, A. C., W, B. Pomeroy, C. E. Martin, and P. Gebhard. 1953. *Sexual Behavior in the Human Female.* Philadelphia: Saunders.

Kirzinger, A., A. Kearney, A. Montero, G. Sparks, L. Dawson, and M. Brodie. 2022, March. *KFF/ The Washington Post Trans Survey.* Washington, DC: The Washington Post/KFF Survey Project. https://files.kff.org/attachment/REPORT-KFF-The-Washington-Post-Trans-Survey.pdf.

Kishida, T., J. Thewissen, T. Hayakawa, H. Imai, and K. Agata. 2015. "Aquatic Adaptation and the Evolution of Smell and Taste in Whales." *Zoological Letters* 1: 9. https://doi.org/10.1186/s40851-014-0002-z.

Kleven, H., C. Landais, and J. Egholt Søgaard. 2018. "Children and Gender Inequality: Evidence from Denmark." National Bureau of Economic Research Working Paper Series No. 24219, Cambridge, MA, 2018. https://doi.org/10.3386/w24219.

Kocot, K. M. 2016. "On 20 Years of Lophotrochozoa." *Organisms Diversity & Evolution* 16: 329–343, https://doi.org/10.1007/s13127-015-0261-3.

Kovacs, C. S., and H. M. Kronenberg. 1997. "Maternal-Fetal Calcium and Bone Metabolism during Pregnancy, Puerperium, and Lactation." *Endocrine Reviews* 18, no. 6: 832–872. https://doi.org/10.1210/edrv.18.6.0319.

Kozlowski, L. T., and J. E. Cutting. 1977. "Recognizing the Sex of a Walker from a Dynamic Point-Light Display." *Perception & Psychophysics* 21, no. 6: 575–580. https://doi.org/https://doi.org/10.3758/BF03198740.

Laine, V. N., T. B. Sackton, and M. Meselson. 2022. "Genomic Signature of Sexual Reproduction in the Bdelloid Rotifer *Macrotrachella quadricornifera*." *Genetics* 220, no. 2: iyab221. https://doi.org/10.1093/genetics/iyab221.

Lake, J. A., E. Henderson, M. Oakes, and M. W. Clark. 1984. "Eocytes: A New Ribosome Structure Indicates a Kingdom with a Close Relationship to Eukaryotes." *Proceedings of the National Academy of Sciences of the United States of America* 81, no. 12: 3786–3790. https://doi.org/10.1073/pnas.81.12.3786.

Laney, S. R., Olson, R. J., and Sosik, H. M. 2012. "Diatoms Favor Their Younger Daughters," *Limnology and Oceanography* 57, no. 5: 1572–78, https://doi:10.4319/lo.2012.57.5.1572.

Le Guin, U. K. 1968. *The Left Hand of Darkness*. New York: Library of America.

Lenz, W. 1988. "A Short History of Thalidomide Embryopathy." *Teratology* 38, no. 3: 203–215. https://doi.org/https://doi.org/10.1002/tera.1420380303.

LeVay, S. 1991. "A Difference in Hypothalamic Structure between Heterosexual and Homosexual Men." *Science* 253, no. 5023: 1034–1037. https://doi.org/10.1126/science.1887219.

Lévi-Strauss, C. 1947. *Les structures élémentaires d la parenté*. Paris: Presses Universitaires de France.

Lewis, G. 2008. "Maternal Mortality in the Developing World: Why Do Mothers Really Die?" *Obstetric Medicine* 1, no. 1: 2–6. https://doi.org/10.1258/om.2008.080019.

Lindberg, L. D., and I. Maddow-Zimet. 2012. "Consequences of Sex Education on Teen and Young Adult Sexual Behaviors and Outcomes." *Journal of Adolescent Health* 51, no. 4: 332–338. https://doi.org/10.1016/j.jadohealth.2011.12.028.

Lindeman, R. E., M. D. Gearhart, A. Minkina, A. D. Krentz, V. J. Bardwell, and D. Zarkower. 2015. "Sexual Cell-Fate Reprogramming in the Ovary by DMRT1." *Current Biology* 25, no. 6: 764–771. https://doi.org/10.1016/j.cub.2015.01.034.

Littman, L. 2018. "Parent Reports of Adolescents and Young Adults Perceived to Show Signs of a Rapid Onset of Gender Dysphoria." *PLoS One* 13, no. 8: e0202330. https://doi.org/10.1371/journal.pone.0202330.

Liu, D.W.C. 2012. "EarthViewer." Howard Hughes Medical Institute. https://media.hhmi.org/biointeractive/earthviewer_web/earthviewer.html. Accessed February, 2024.

Liu, K. A. and N. A. DiPietro Mager. 2016. "Women's Involvement in Clinical Trials: Historical Perspective and Future Implications" *Pharmacy Practice* 14, no. 1 (January-March, 2016): 708. https://doi.org/https://doi.org/10.18549/PharmPract.2016.01.708.

Lockett, E., A. Noe, R.F. Gasser, R.J. Cork, S. Blanchard, T. Monroe, C. Thouron, Z. St. Onge, A. Chang, K. Mayur, M. Shaw, E. Wilson, and W. F. Discher. 2011. "The Virtual Human Embryo

Project." Concord, NH: Endowment for Human Development. https://www.ehd.org/virtual
-human-embryo/.

Lofquist, D. 2011. "Same-Sex Couple Households." In *American Community Survey Briefs*, ed. U.S.
Census Bureau. Washington, DC: U.S. Department of Commerce.

Luo, Z. X., C. X. Yuan, Q. J. Meng, and Q. Ji. 2011. "A Jurassic Eutherian Mammal and Divergence of Marsupials and Placentals." *Nature* 476, no. 7361: 442–445. https://doi.org/10.1038
/nature10291.

Maas, A. H. E. M., and Y. E. A. Appelman. 2010. "Gender Differences in Coronary Heart Disease." *Netherlands Heart Journal* 18, no. 12: 598–603. https://doi.org/10.1007/s12471-010-0841-y.

Macdonald, J. D. 1869. "On the Structure of the Diatomaceous Frustule, and Its Genetic Cycle."
Annals and Magazine of Natural History 3, no. 13: 1–8.

Mace, Terrence R. 2000. "Time Budget and Pair-Bond Dynamics in the Comb-Crested Jacana
Irediparra Gallinacea: A Test of Hypotheses." *Emu* 100, no. 1: 31–41. https://doi.org/10.1071
/MU9844.

Mackay-Sim, A., D. Duvall, and B. M. Graves. 1985. "The West Indian Manatee (*Trichechus manatus*) Lacks a Vomeronasal Organ." *Brain, Behavior, and Evolution* 27, no. 2–4: 186–194. https://
doi.org/10.1159/000118729.

Mackiewicz, Mark, Andrey Tatarenkov, Andrew Perry, J. Ryce Martin, John F. Elder Jr., David L.
Bechler, and John C. Avise. 2006. "Microsatellite Documentation of Male-Mediated Outcrossing between Inbred Laboratory Strains of the Self-Fertilizing Mangrove Killifish (*Kryptolebias
marmoratus*)." *Journal of Heredity* 97, no. 5: 508–513. https://doi.org/10.1093/jhered/esl017.

Mackiewicz, Mark, Andrey Tatarenkov, D. Scott Taylor, Bruce J. Turner, and John C. Avise.
2006. "Extensive Outcrossing and Androdioecy in a Vertebrate Species That Otherwise
Reproduces as a Self-Fertilizing Hermaphrodite." *Proceedings of the National Academy of Sciences* 103, no. 26: 9924–9928. https://doi.org/doi:10.1073/pnas.0603847103.

Maienschein, J., M. Glitz, and G. E. Allen. 2004. *The Department of Embryology* Vol. 5. *Centennial
History of the Carnegie Institution of Washington*. Cambridge: Cambridge University Press.

Marletaz, F., Ktca Peijnenburg, T. Goto, N. Satoh, and D. S. Rokhsar. 2019. "A New Spiralian Phylogeny Places the Enigmatic Arrow Worms among Gnathiferans," *Current Biology* 29, no. 2: 312–318.

Marques, Antonio C., and Allen G. Collins. 2004. "Cladistic Analysis of Medusozoa and Cnidarian Evolution." *Invertebrate Biology* 123, no. 1: 23–42. https://doi.org/https://doi.org/10.1111
/j.1744-7410.2004.tb00139.x.

Marshall, C. R., E. C. Raff, and R. A. Raff. 1994. "Dollo's Law and the Death and Resurrection
of Genes." *Proceedings of the National Academy of Sciences of the United States of America* 91, no.
25: 12283–12287. https://doi.org/10.1073/pnas.91.25.12283.

Martin, W., and M. Müller. 1998. "The Hydrogen Hypothesis for the First Eukaryote." *Nature*
392, no. 6671: 37–41. https://doi.org/10.1038/32096.

Matson, C. K., M. W. Murphy, A. L. Sarver, M. D. Griswold, V. J. Bardwell, and D. Zarkower.
2011. "DMRT1 Prevents Female Reprogramming in the Postnatal Mammalian Testis." *Nature*
476, no. 7358: 101–104. https://doi.org/10.1038/nature10239.

McLean, C. Y., P. L. Reno, A. A. Pollen, A. I. Bassan, T. D. Capellini, C. Guenther, V. B. Indjeian,
X. Lim, D. B. Menke, B. T. Schaar, A. M. Wenger, G. Bejerano, and D. M. Kingsley. 2011.
"Human-Specific Loss of Regulatory DNA and the Evolution of Human-Specific Traits."
Nature 471, no. 7337: 216–219. https://doi.org/10.1038/nature09774.

Mead, H. M. 2003. *Tikanga Maori: Living by Maori Values* Wellington, New Zealand: Huia.

Meischke, Hendrika, Mary Pat Larsen, and Mickey S. Eisenberg. 1998. "Gender Differences in Reported Symptoms for Acute Myocardial Infarction: Impact on Prehospital Delay Time Interval." *American Journal of Emergency Medicine* 16, no. 4: 363–366. https://doi.org/https://doi.org/10.1016/S0735-6757(98)90128-0.

Meng, Q. J., Q. Ji, Y. G. Zhang, D. Liu, D. M. Grossnickle, and Z. X. Luo. 2015. "Mammalian Evolution: An Arboreal Docodont from the Jurassic and Mammaliaform Ecological Diversification." *Science* 347, no. 6223: 764–768. https://doi.org/10.1126/science.1260879.

Mettler, K. 2019. "Anna the Anaconda Got Pregnant All by Herself—by 'Virgin Birth.' " *Washington Post*, May 24, 2019, Science.

Mi, S., X. Lee, X. Li, G. M. Veldman, H. Finnerty, E. LaVallie, X. Y. Tang, P. Edouard, S. Howes, J. C. Keith Jr., and J. M. McCoy. 2000. "Syncytin Is a Captive Retroviral Envelope Protein Involved in Human Placental Morphogenesis." *Nature* 403, no. 6771: 785–789. https://doi.org/10.1038/35001608.

Michiels, N. K., and L. J. Newman. 1998. "Sex and Violence in Hermaphrodites." *Nature* 391, no. 6668: 647–647. https://doi.org/10.1038/35527.

Michod, R. E., and B. R. Levin, eds. 1988. *The Evolution of Sex: An Examination of Current Ideas.* Sunderland, MA: Sinauer.

Mikkola, M., ed. 2023, Spring. *Feminist Perspectives on Sex and Gender.* Ed. E. N. Zalta and U. Nodelman. *The Stanford Encyclopedia of Philosophy* Palo Alto: Stanford University, Department of Philosophy.

Mitteroecker, P., S. M. Huttegger, B. Fischer, and M. Pavlicev. 2016. "Cliff-Edge Model of Obstetric Selection in Humans." *Proceedings of the National Academy of Sciences of the United States of America* 113, no. 51: 14680–14685. https://doi.org/10.1073/pnas.1612410113.

Moczek, A. P., and D. J. Emlen. 1999. "Proximate Determination of Male Horn Dimorphism in the Beetle *Onthophagus taurus* (Coleoptera: Scarabaeidae)." *Journal of Evolutionary Biology* 12, no. 1: 27–37. https://doi.org/10.1046/j.1420-9101.1999.00004.x.

Morgan, T. H. 1903. "Recent Theories in Regard to the Determination of Sex." *Popular Science Monthly* 64 (December): 97–116.

Morgan, T. H. 1910. "Sex Limited Inheritance in *Drosophila*." *Science* 32, no. 812: 120–122. https://doi.org/10.1126/science.32.812.120.

Mori, T., H. Kuroiwa, T. Higashiyama, and T. Kuroiwa. 2006. "GENERATIVE CELL SPECIFIC 1 Is Essential for Angiosperm Fertilization." *Nature Cell Biology Nature Cell Biology* 8, no. 1: 64–71. https://doi.org/10.1038/ncb1345.

Morran, L. T., M. D. Parmenter, and P. C. Phillips. 2009. "Mutation Load and Rapid Adaptation Favour Outcrossing over Self-Fertilization." *Nature* 462, no. 7271: 350–352.

Morran, L. T., O. G. Schmidt, I. A. Gelarden, R. C. Parrish II, and C. M. Lively. 2011. "Running with the Red Queen: Host-Parasite Coevolution Selects for Biparental Sex." *Science* 333, no. 6039: 216–218. https://doi.org/10.1126/science.1206360.

Nagamine, Claude M. 1996. "Sex Reversal in Mammals." In *Advances in Genome Biology*, ed. Ram S. Verma, 53–118. Atlanta, GA: JAI.

Naguib, S. 2010. "Horizons and Limitations of Muslim Feminist Hermeneutics: Reflections on the Menstruation Verse." In *New Topics in Feminist Philosophy of Religion*, ed. P. S. Anderson, 33–49. New York: Springer.

Nasrallah, M. E., P. Liu, and J. B. Nasrallah. 2002. "Generation of Self-Incompatible *Arabidopsis thaliana* by Transfer of Two S Locus Genes from *A. lyrata.*" *Science* 297, no. 5579: 247–249. https://doi.org/10.1126/science.1072205.

Nass, S., and M. M. Nass. 1963. "Intramitochondrial Fibers with DNA Characteristics. II. Enzymatic and Other Hydrolytic Treatments." *Journal of Cell Biology* 19: 613–629. https://doi.org/10.1083/jcb.19.3.613.

New York Magazine editors. 2017. "What We Learned about Sexual Desire from 10 Years of Pornhub User Data." *New York Magazine—The Cut,* June 11.

Nicolas, M. 1972. *Croyances et pratiques populaires turques concernant les naissances.* Paris: Publications Orientalistes de France.

Office of Juvenile Justice and Delinquency Prevention. 2021. "Easy Access to the FBI's Supplementary Homicide Reports (EZASHR)." Pittsburgh, PA: National Center for Juvenile Justice. https://www.ojjdp.gov/ojstatbb/ezashr/.

Okada, S., and L. D. Peachy. 1957. "Effect of Gamma Irradiation on the Desoxyribonuclease II Activity of Isolated Mitochondria." *Journal of Biophysical and Biochemical Cytology* 3, no. 2: 239–248. https://doi.org/10.1083/jcb.3.2.239.

Oyewumi, O. 1997. *The Invention of Women.* Minneapolis: University of Minnesota Press.

Paczolt, K. A., and A. G. Jones. 2010. "Post-Copulatory Sexual Selection and Sexual Conflict in the Evolution of Male Pregnancy." *Nature* 464, no. 7287: 401–404. https://doi.org/10.1038/nature08861.

Page, D. C., E. M. Fisher, B. McGillivray, and L. G. Brown. 1990. "Additional Deletion in Sex-Determining Region of Human Y Chromosome Resolves Paradox of X,t(Y;22) Female." *Nature* 346, no. 6281: 279–281. https://doi.org/10.1038/346279a0.

Page, D. C., R. Mosher, E. M. Simpson, E. M. Fisher, G. Mardon, J. Pollack, B. McGillivray, A. de la Chapelle, and L. G. Brown. 1987. "The Sex-Determining Region of the Human Y Chromosome Encodes a Finger Protein." *Cell* 51, no. 6: 1091–1104. https://doi.org/10.1016/0092-8674(87)90595-2.

Painter, T. S. 1921. "The Y-Chromosome in Mammals." *Science* 53, no. 1378: 503–504. https://doi.org/10.1126/science.53.1378.503.

Pandian, T. J. 2019. *Reproduction and Development in Annelida.* Boca Raton, FL: CRC Press.

Panheleux, M., M. Bain, M. S. Fernandez, I. Morales, J. Gautron, J. L. Arias, S. E. Solomon, M. Hincke, and Y. Nys. 1999. "Organic Matrix Composition and Ultrastructure of Eggshell: A Comparative Study." *British Poultry Science* 40, no. 2: 240–252. https://doi.org/10.1080/00071669987665.

Parent, H., and M. Zatoń. 2016. "Sexual Dimorphism in the Bathonian Morphoceratid Ammonite *Polysphinctites tenuiplicatus.*" *Acta Palaeontologica Polonica* 61, no. 4: 875–884. https://doi.org/10.4202/app.00261.2016.

Parisi, M., R. Nuttall, D. Naiman, G. Bouffard, J. Malley, J. Andrews, S. Eastman, and B. Oliver. 2003. "Paucity of Genes on the *Drosophila* X Chromosome Showing Male-Biased Expression." *Science* 299, no. 5607: 697–700. https://doi.org/10.1126/science.1079190.

Parker, G. A. 1970. "Sperm Competition and Its Evolutionary Consequences in the Insects." *Biological Reviews* 45, no. 4: 525–567. https://doi.org/https://doi.org/10.1111/j.1469-185X.1970.tb01176.x.

Parker, G. A., R. R. Baker, and V. G. Smith. 1972. "The Origin and Evolution of Gamete Dimorphism and the Male-Female Phenomenon." *Journal of Theoretical Biology* 36, no. 3: 529–553. https://doi.org/10.1016/0022-5193(72)90007-0.

Parkes, A., V. Strange, D. Wight, C. Bonell, A. Copas, M. Henderson, K. Buston, J. Stephenson, A. Johnson, E. Allen, and G. Hart. 2011. "Comparison of Teenagers' Early Same-Sex and Heterosexual Behavior: UK Data from the SHARE and RIPPLE studies." *Journal of Adolescent Health* 48, no. 1: 27–35. https://doi.org/10.1016/j.jadohealth.2010.05.010.

Pavelka, Mary S. M., and Linda Marie Fedigan. 1991. "Menopause: A Comparative Life History Perspective." *American Journal of Physical Anthropology* 34, no. S13 (American Journal of Physical Anthropology Annual Meeting Issue): 13–38. https://doi.org/https://doi.org/10.1002/ajpa.1330340604.

Pavlicev, M., R. Romero, and P. Mitteroecker. 2020. "Evolution of the Human Pelvis and Obstructed Labor: New Explanations of an Old Obstetrical Dilemma." *American Journal of Obstetrics & Gynecology* 222, no. 1: 3–16. https://doi.org/10.1016/j.ajog.2019.06.043.

Pavlicev, M., and G. Wagner. 2016. "The Evolutionary Origin of Female Orgasm." *Journal of Experimental Zoology Part B: Molecular and Developmental Evolution* 326, no. 6: 326–337. https://doi.org/10.1002/jez.b.22690.

Pavlicev, M., A. M. Zupan, A. Barry, S. Walters, K. M. Milano, H. J. Kliman, and G. P. Wagner. 2019. "An Experimental Test of the Ovulatory Homolog Model of Female Orgasm." *Proceedings of the National Academy of Sciences of the United States of America* 116, no. 41: 20267–20273. https://doi.org/10.1073/pnas.1910295116.

Perelman, P., W. E. Johnson, C. Roos, H. N. Seuanez, J. E. Horvath, M. A. Moreira, B. Kessing, J. Pontius, M. Roelke, Y. Rumpler, M. P. Schneider, A. Silva, S. J. O'Brien, and J. Pecon-Slattery. 2011. "A Molecular Phylogeny of Living Primates." *PLoS Genetics* 7, no. 3: e1001342. https://doi.org/10.1371/journal.pgen.1001342.

Peters, Sanne A. E., Paul Muntner, and Mark Woodward. 2019. "Sex Differences in the Prevalence of, and Trends in, Cardiovascular Risk Factors, Treatment, and Control in the United States, 2001 to 2016." *Circulation* 139, no. 8: 1025–1035. https://doi.org/doi:10.1161/CIRCULATIONAHA.118.035550.

Petersen, C. W. 1995. "Reproductive Behavior, Egg Trading, and Correlates of Male Mating Success in the Simultaneous Hermaphrodite, *Serranus tabacarius*." *Environmental Biology of Fishes* 43, no. 4: 351–361. https://doi.org/10.1007/BF00001169.

Petersen, J. K. 2021. *A Comprehensive Guide to Intersex*. Philadephia: Jessica Kingsley.

Pfitzer, E. 1871. "Untersuchungen über Bau und Entwickelung der Bacillariaceen (Diatomaceen)." *Botanische Abhandlungen aus dem Gebiet der Morphologie und Physiologie*. 1, no. 2: 1–189.

Phipps, E. A., R. Thadhani, T. Benzing, and S. A. Karumanchi. 2019. "Pre-Eclampsia: Pathogenesis, Novel Diagnostics and Therapies." *Nature Reviews Nephrology* 15, no. 5: 275–289. https://doi.org/10.1038/s41581-019-0119-6.

Pianka, H. D. 1974. "Ctenophora." In *Reproduction of Marine Invertebrates: Acoelomate and Pseudocoelomate Metazoans*, ed. A. C. Giese and J. S. Pearse, 201–265. New York: Academic Press.

Pomiankowski, A, and W. D. Hamilton. 1987. "Sexual Selection: The Handicap Principle Does Work—Sometimes." *Proceedings of the Royal Society of London, Series B: Biological Sciences* 231, no. 1262: 123–145. https://doi.org/doi:10.1098/rspb.1987.0038.

Profet, M. 1993. "Menstruation as a Defense against Pathogens Transported by Sperm." *Quarterly Review of Biology* 68, no. 3: 335–386. https://doi.org/10.1086/418170.

Qu, M., S. Ding, M. Schartl, and M. C. Adolfi. 2020. "Spatial and Temporal Expression Pattern of Sex-Related Genes in Ovo-Testis of the Self-Fertilizing Mangrove Killifish (*Kryptolebias marmoratus*)." *Gene* 742: 144581. https://doi.org/10.1016/j.gene.2020.144581.

Raff, R. A. 1996. *The Shape of Life*. Chicago: University of Chicago Press.

Rajpert-De Meyts, E., N. Jorgensen, N. Graem, J. Muller, R. L. Cate, and N. E. Skakkebaek. 1999. "Expression of Anti-Mullerian Hormone during Normal and Pathological Gonadal Development: Association with Differentiation of Sertoli and Granulosa Cells." *Journal of Clinical Endocrinology and Metabolism* 84, no. 10: 3836–3844. https://doi.org/10.1210/jcem.84.10.6047.

Ramm, Steven A., Aline Schlatter, Maude Poirier, and Lukas Schärer. 2015. "Hypodermic Self-Insemination as a Reproductive Assurance Strategy." *Proceedings of the Royal Society of London, Series B: Biological Sciences* 282, no. 1811: 20150660. https://doi.org/doi:10.1098/rspb.2015.0660.

Raymond, C. S., C. E. Shamu, M. M. Shen, K. J. Seifert, B. Hirsch, J. Hodgkin, and D. Zarkower. 1998. "Evidence for Evolutionary Conservation of Sex-Determining Genes." *Nature* 391: 873–881.

Reed, M. B., P. A. Handschuh, M. Klöbl, M. E. Konadu, U. Kaufmann, A. Hahn, G. S. Kranz, M. Spies, and R. Lanzenberger. 2023. "The Influence of Sex Steroid Treatment on Insular Connectivity in Gender Dysphoria." *Psychoneuroendocrinology* 155: 106336. https://doi.org/https://doi.org/10.1016/j.psyneuen.2023.106336.

Reinke, V., I. S. Gil, S. Ward, and K. Kazmer. 2004. "Genome-Wide Germline-Enriched and Sex-Biased Expression Profiles in *Caenorhabditis elegans*." *Development* 131, no. 2: 311–323.

Rennison, S. M. 2002. *Rape and Sexual Assault: Reporting to Police and Medical Attention, 1992–2000*. Washington, DC: U.S. Department of Justice https://bjs.ojp.gov/content/pub/pdf/rsarp00.pdf.

Restar, Arjee Javellana. 2020. "Methodological Critique of Littman's (2018) Parental-Respondents Accounts of 'Rapid-Onset Gender Dysphoria.'" *Archives of Sexual Behavior* 49, no. 1: 61–66. https://doi.org/10.1007/s10508-019-1453-2.

Rice, W. R., U. Friberg, and S. Gavrilets. 2012. "Homosexuality as a Consequence of Epigenetically Canalized Sexual Development." *Quarterly Review of Biology* 87, no. 4: 343–368. https://doi.org/10.1086/668167.

Richardson, S. 2013. *Sex Itself: The Search for Male and Female in the Human Genome* Chicago: University of Chicago Press.

Richerson, P., R. Baldini, A. V. Bell, K. Demps, K. Frost, V. Hillis, S. Mathew, E. K. Newton, N. Naar, L. Newson, C. Ross, P. E. Smaldino, T. M. Waring, and M. Zefferman. 2016. "Cultural Group Selection Plays an Essential Role in Explaining Human Cooperation: A Sketch of the Evidence." *Behavioral and Brain Sciences* 39: e30. https://doi.org/10.1017/S0140525X1400106X.

Riesgo, Ana, Nathan Farrar, Pamela J. Windsor, Gonzalo Giribet, and Sally P. Leys. 2014. "The Analysis of Eight Transcriptomes from All Poriferan Classes Reveals Surprising Genetic Complexity in Sponges." *Molecular Biology and Evolution* 31, no. 5: 1102–1120. https://doi.org/10.1093/molbev/msu057.

Ritchison, G. 2024. "XXIII—Avian Reproduction III: Clutch Size, Incubation, & Hatching." *Avian Biology*. https://avesbiology.com2024.

Ross, C., and T. E. Boroviak. 2020. "Origin and Function of the Yolk Sac in Primate Embryogenesis." *Nature Communications* 11, no. 1: 3760. https://doi.org/10.1038/s41467-020-17575-w.

Rucker, P. 2015. "Trump Says Fox's Megyn Kelly Had 'Blood Coming out of Her Wherever.'" *Washington Post* Weblog post, August 8, 2015, Politics. https://www.washingtonpost.com/news/post-politics/wp/2015/08/07/trump-says-foxs-megyn-kelly-had-blood-coming-out-of-her-wherever/.

Ryabko, D., and Z. Reznikova. 2015. "On the Evolutionary Origins of Differences in Sexual Preferences." *Frontiers in Psychology* 6: 628. https://doi.org/10.3389/fpsyg.2015.00628.

Sadro, S. 2001. "Porifera: The Sponges." In *An Identification Guide to Larval Marine Invertebrates of the Pacific Northwest*, ed. A. L. Shanks, 5–12. Corvallis: Oregon State University Press.

Sagan, L. 1967. "On the Origin of Mitosing Cells." *Journal of Theoretical Biology* 14, no. 3: 255–274. https://doi.org/10.1016/0022-5193(67)90079-3.

Saisuwan, P. 2015. "Kathoey and the Linguistic Construction of Gender Identity in Thailand." In *Language, Sexuality, and Power: Studies in Intersectional Sociolinguistics, Studies in Language Gender and Sexuality*, ed. E. Levon and R. Beline Mendes, 189–214. New York: Oxford Academic.

Samolovitch, N. 2023. "Brilliant Women in U.S. Horse Racing History." *America's Best Racing* https://www.americasbestracing.net/the-sport/2023-brilliant-women-us-horse-racing-history.

Sanders, A. R., G. W. Beecham, S. Guo, K. Dawood, G. Rieger, J. A. Badner, E. S. Gershon, R. S. Krishnappa, A. B. Kolundzija, J. Duan, M. G. S. Collaboration, P. V. Gejman, J. M. Bailey, and E. R. Martin. 2017. "Genome-Wide Association Study of Male Sexual Orientation." *Scientific Reports* 7, no. 1: 16950. https://doi.org/10.1038/s41598-017-15736-4.

Schlupp, I. 2021. *Male Choice, Female Competition, and Female Ornaments* New York: Oxford University Press.

Schofield, R. 1921. "Inheritance of Webbed Toes." *Journal of Heredity* 12, no. 9: 400–401.

Schurman, J. G. 1893. *The Eithical Import of Darwinism*. 2nd ed. New York: Charles Scribner's Sons.

Schwartz, G., R. M. Kim, A. B. Kolundzija, G. Rieger, and A. R. Sanders. 2010. "Biodemographic and Physical Correlates of Sexual Orientation in Men." *Archives of Sexual Behavior* 39, no. 1: 93–109. https://doi.org/10.1007/s10508-009-9499-1.

Sekizawa, A., S. Satoko, T. Masakazu, S. Sakiko, and N. Yasuhiro. 2013. "Disposable Penis and its Replenishment in a Simultaneous Hermaphrodite," *Biology Letters* 9, no. 2. http://doi.org/10.1098/rsbl.2012.1150.

Sharma, V., T. Lehmann, H. Stuckas, L. Funke, and M. Hiller. 2018. "Loss of RXFP2 and INSL3 Genes in Afrotheria Shows That Testicular Descent is the Ancestral Condition in Placental Mammals." *PLoS Biology* 16, no. 6: e2005293. https://doi.org/10.1371/journal.pbio.2005293.

Shefi, S., P. E. Tarapore, T. J. Walsh, M. Croughan, and P. J. Turek. 2007. "Wet Heat Exposure: A Potentially Reversible Cause of Low Semen Quality in Infertile Men." *International Brazilian Journal of Urology* 33, no. 1: 50–56; discussion 56–57. https://doi.org/10.1590/s1677-55382007000100008.

Shen, Joel, Gerald R. Cunha, Adriane Sinclair, Mei Cao, Dylan Isaacson, and Laurence Baskin. 2018. "Macroscopic Whole-Mounts of the Developing Human Fetal Urogenital-Genital Tract: Indifferent Stage to Male and Female Differentiation." *Differentiation* 103: 5–13. https://doi.org/https://doi.org/10.1016/j.diff.2018.08.003.

Shine, R. 1983. "Reptilian Viviparity in Cold Climates: Testing the Assumptions of an Evolutionary Hypothesis." *Oecologia* 57, no. 3: 397–405. https://doi.org/10.1007/BF00377186.

Shine, R. 1995. "A New Hypothesis for the Evolution of Viviparity in Reptiles." *The American Naturalist* 145, no. 5: 809–823. https://doi.org/10.1086/285769.

Shine, R. 2014. "Evolution of an Evolutionary Hypothesis: A History of Changing Ideas about the Adaptive Significance of Viviparity in Reptiles." *Journal of Herpetology* 48, no. 2: 147–161, 15.

Sikes, J. M., and A. E. Bely. 2008. "Radical Modification of the A-P Axis and the Evolution of Asexual Reproduction in *Convolutriloba* Acoels." *Evolution & Development* 10, no. 5: 619–631, https://doi.org/10.1111/j.1525-142X.2008.00276.x.

Silverman, E. 2022. "Three Virginia Teens Fight 'Period Poverty.'" *Washington Post*, December 10, 2022.

Simion, P., H. Philippe, D. Baurain, M. Jager, D. J. Richter, A. Di Franco, B. Roure, et al. 2017. "A Large and Consistent Phylogenomic Dataset Supports Sponges as the Sister Group to All Other Animals," *Current Biology* 27, no. 7: 958–967. https://doi.org/10.1016/j.cub.2017.02.031.

Sinclair, A. H., P. Berta, M. S. Palmer, J. R. Hawkins, B. L. Griffiths, M. J. Smith, J. W. Foster, A. M. Frischauf, R. Lovell-Badge, and P. N. Goodfellow. 1990. "A Gene from the Human Sex-Determining Region Encodes a Protein with Homology to a Conserved DNA-Binding Motif." *Nature* 346, no. 6281: 240–244. https://doi.org/10.1038/346240a0.

Singh, L., and K. W. Jones. 1982. "Sex Reversal in the Mouse (*Mus musculus*) Is Caused by a Recurrent Nonreciprocal Crossover Involving the X and an Aberrant Y Chromosome." *Cell* 28, no. 2: 205–216. https://doi.org/10.1016/0092-8674(82)90338-5.

Smith, K. 2015. "Placental evolution in therian mammals." In *Great Transformations in Vertebrate Evolution*, edited by K.P. Dial, N. Shubin and E.L. Brainerd, 205-226. Chicago, IL: University of Chicago Press.

Solomon-Lane, T. K., E. J. Crespi, and M. S. Grober. 2013. "Stress and Serial Adult Metamorphosis: Multiple Roles for the Stress Axis in Socially Regulated Sex Change." *Frontiers in Neuroscience* 7: 210. https://doi.org/10.3389/fnins.2013.00210.

Spang, A., J. H. Saw, S. L. Jorgensen, K. Zaremba-Niedzwiedzka, J. Martijn, A. E. Lind, R. van Eijk, C. Schleper, L. Guy, and T. J. G. Ettema. 2015. "Complex Archaea That Bridge the Gap between Prokaryotes and Eukaryotes." *Nature* 521, no. 7551: 173–179. https://doi.org/10.1038/nature14447.

Speijer, D. 2017. "Alternating Terminal Electron-Acceptors at the Basis of Symbiogenesis: How Oxygen Ignited Eukaryotic Evolution." *Bioessays* 39, no. 2: 1600174. https://doi.org/10.1002/bies.201600174.

Spurway, J., P. Logan, and S. Pak. 2012. "The Development, Structure and Blood Flow within the Umbilical Cord with Particular Reference to the Venous System." *Australasian Journal of Ultrasound in Medicine* 15, no. 3: 97–102. https://doi.org/10.1002/j.2205-0140.2012.tb00013.x.

Stansfield, E., K. Kumar, P. Mitteroecker, and N. D. S. Grunstra. 2021. "Biomechanical Trade-offs in the Pelvic Floor Constrain the Evolution of the Human Birth Canal." *Proceedings of the National Academy of Sciences of the United States of America* 118, no. 16: e2022159118. https://doi.org/10.1073/pnas.2022159118.

Starkweather, Katherine E., and Raymond Hames. 2012. "A Survey of Non-Classical Polyandry." *Human Nature* 23, no. 2: 149–172. https://doi.org/10.1007/s12110-012-9144-x.

Stein, K., E. Prondvai, T. Huang, J. M. Baele, P. M. Sander, and R. Reisz. 2019. "Structure and Evolutionary Implications of the Earliest (Sinemurian, Early Jurassic) Dinosaur Eggs and Eggshells." *Scientific Reports* 9, no. 1: 4424. https://doi.org/10.1038/s41598-019-40604-8.

Stemple, Lara, Andrew Flores, and Ilan H. Meyer. 2017. "Sexual Victimization Perpetrated by Women: Federal Data Reveal Surprising Prevalence." *Aggression and Violent Behavior* 34: 302–311. https://doi.org/https://doi.org/10.1016/j.avb.2016.09.007.

Stern, C. 1957. "The Problem of Complete Y-Linkage in Man." *American Journal of Human Genetics* 9, no. 3: 147–166.

Stevens, N. M. 1905. *Studies in Spermatogenesis with Especial Reference to the "Accessory Chromosome."* Washington, DC: Carnegie Institution of Washington.

Stott, R. 2003. *Darwin and the Barnacle.* New York: Norton.

Stowers, L., T. E. Holy, M. Meister, C. Dulac, and G. Koentges. 2002. "Loss of Sex Discrimination and Male-Male Aggression in Mice Deficient for TRP2." *Science* 295, no. 5559: 1493–1500. https://doi.org/10.1126/science.1069259.

Strassmann, B. I. 1996. "The evolution of endometrial cycles and menstruation." *Quarterly Review of Biology* 71 (2): 181-220. https://doi.org/10.1086/419369.

Stripe, N. 2021. "Appendix Tables: Homicide in England and Wales." Ed. Office for National Statistics. United Kingdom. https://www.ons.gov.uk/peoplepopulationandcommunity/crimeandjustice/datasets/appendixtableshomicideinenglandandwales.

Stryker, S. 2017. *Transgender History.* 2nd ed. New York: Seal Press.

Sturtevant, A. H. 1913. "The Linear Arrangement of Six Sex-Linked Factors in Drosophila, as Shown by Their Mode of Association." *Journal of Experimental Zoology* 14, no. 1: 43–59. https://doi.org/https://doi.org/10.1002/jez.1400140104.

Sykes, B. 2004. *Adam's Curse: A Future Without Men.* New York: Norton.

Takatsu, K., K. Miyaoku, S. R. Roy, Y. Murono, T. Sago, H. Itagaki, M. Nakamura, and T. Tokumoto. 2013. "Induction of Female-to-Male Sex Change in Adult Zebrafish by Aromatase Inhibitor Treatment." *Scientific Reports* 3: 3400. https://doi.org/10.1038/srep03400.

Teklenburg, G., M. Salker, M. Molokhia, S. Lavery, G. Trew, T. Aojanepong, H. J. Mardon, A. U. Lokugamage, R. Rai, C. Landles, B. A. Roelen, S. Quenby, E. W. Kuijk, A. Kavelaars, C. J. Heijnen, L. Regan, J. J. Brosens, and N. S. Macklon. 2010. "Natural Selection of Human Embryos: Decidualizing Endometrial Stromal Cells Serve as Sensors of Embryo Quality upon Implantation." *PLoS One* 5, no. 4: e10258. https://doi.org/10.1371/journal.pone.0010258.

Tentelier, C., O. Lepais, N. Larranaga, A. Manicki, F. Lange, and J. Rives. 2016. "Sexual Selection Leads to a Tenfold Difference in Reproductive Success of Alternative Reproductive Tactics in Male Atlantic Salmon." *Naturwissenschaften* 103, no. 5–6: 47. https://doi.org/10.1007/s00114-016-1372-1.

Thomas, C. G., R. Li, H. E. Smith, G. C. Woodruff, B. Oliver, and E. S. Haag. 2012. "Simplification and Desexualization of Gene Expression in Self-Fertile Nematodes." *Current Biology* 22, no. 22: 2167–72. https://doi.org/10.1016/j.cub.2012.09.038.

Thomas, C. G., G. C. Woodruff, and E. S. Haag. 2012. "Causes and Consequences of the Evolution of Reproductive Mode in Caenorhabditis Nematodes." *Trends in Genetics* 28, no. 5: 213–220. https://doi.org/10.1016/j.tig.2012.02.007.

Thomas, J., K. Handasyde, M. L. Parrott, and P. Temple-Smith. 2017. "The Pplatypus Nest: Burrow Structure and Nesting Behaviour in Captivity." *Australian Journal of Zoology* 65, no. 6: 347–356. https://doi.org/https://doi.org/10.1071/ZO18007.

Thyen, Ute, Kathrin Lanz, Paul-Martin Holterhus, and Olaf Hiort. 2006. "Epidemiology and Initial Management of Ambiguous Genitalia at Birth in Germany." *Hormone Research* 66, no. 4: 195–203. https://doi.org/10.1159/000094782.

Tian, X., J. Gautron, P. Monget, and G. Pascal. 2010. "What Makes an Egg Unique? Clues from Evolutionary Scenarios of Egg-Specific Genes." *Biology of Reproduction* 83, no. 6: 893–900. https://doi.org/10.1095/biolreprod.110.085019.

Tobias, J. A., R. Montgomerie, and B. E. Lyon. 2012. "The Evolution of Female Ornaments and Weaponry: Social Selection, Sexual Selection and Ecological Competition." *Philosophical Transactions of the Royal Society of London, Series B: Biological Sciences* 367, no. 1600: 2274–2293. https://doi.org/10.1098/rstb.2011.0280.

Todd, B. K., R. A. Fischer, S. Di Costa, A. Roestorf, K. Harbour, P. Hardiman, and J. A. Barry. 2018. "Sex Differences in Children's Toy Preferences: A Systematic Review, Meta-Regression, and Meta-Analysis." *Infant and Child Development* 27, no. 2: e2064. https://doi.org/https://doi.org/10.1002/icd.2064.

Todd, E. V., O. Ortega-Recalde, H. Liu, M. S. Lamm, K. M. Rutherford, H. Cross, M. A. Black, O. Kardailsky, J. A. Marshall Graves, T. A. Hore, J. R. Godwin, and N. J. Gemmell. 2019. "Stress, Novel Sex Genes, and Epigenetic Reprogramming Orchestrate Socially Controlled Sex Change." *Science Advances* 5, no. 7: eaaw7006. https://doi.org/10.1126/sciadv.aaw7006.

Trinkaus, E., and A. P. Buzhilova. 2012. "The Death and Burial of Sunghir 1." *International Journal of Osteoarchaeology* 22, no. 6: 655–666. https://doi.org/https://doi.org/10.1002/oa.1227.

Trotier, D., C. Eloit, M. Wassef, G. Talmain, J. L. Bensimon, K. B. Doving, and J. Ferrand. 2000. "The Vomeronasal Cavity in Adult Humans." *Chemical Senses* 25, no. 4: 369–380. https://doi.org/10.1093/chemse/25.4.369.

True, J. R., and E. S. Haag. 2001. "Developmental System Drift and Flexibility in Evolutionary Trajectories." *Evolution & Development* 3, no. 2: 109–119.

Tschantret, Joshua. 2020. "Revolutionary Homophobia: Explaining State Repression against Sexual Minorities." *British Journal of Political Science* 50, no. 4: 1459–1480. https://doi.org/10.1017/S0007123418000480.

Umen, J. G. 2014. "Green Algae and the Origins of Multicellularity in the Plant Kingdom." *Cold Spring Harbor Perspectives in Biology* 6, no. 11: a016170. https://doi.org/10.1101/cshperspect.a016170.

UN Office of the High Commissioner for Human Rights. 2015. *Discrimination and Violence against Individuals Based on Their Sexual Orientation and Gender Identity* . Geneva: United Nationa. https://digitallibrary.un.org/record/797193/files/A_HRC_29_23-EN.pdf.

United Nations Human Rights Council. 2010. *Report of the Office of the United Nations High Commissioner for Human Rights on Preventable Maternal Mortality and Morbidity and Human Rights.* Geneva: United Nations. https://www.ohchr.org/sites/default/files/Documents/Issues/Women/WRGS/Health/ReportMaternalMortality.pdf.

U.S. Agency for International Development. 2023. *The Empowerment of Women through Sanitation, and Hygiene Programs, FY 2022.* USAID Report to Congress (Washington, DC, March 2, 2023). https://www.usaid.gov/reports/empowerment-women-sanitation-hygiene/fy-2022; U.S. Census Bureau. 2021. "2015 ACS 1-Year Estimates." Washington, DC: U.S. Government Publishing Office.

United States Sentencing Commission. 2018. "Quick Facts: Sexual Abuse Offenders." Washington, DC: United States Sentencing Commission, https://www.ussc.gov/sites/default/files/pdf/research-and-publications/quick-facts/Sexual_Abuse_FY18.pdf.

Vakhrusheva, O. A., E. A. Mnatsakanova, Y. R. Galimov, T. V. Neretina, E. S. Gerasimov, S. A. Naumenko, S. G. Ozerova, A. O. Zalevsky, I. A. Yushenova, F. Rodriguez, I. R. Arkhipova, A. A. Penin, M. D. Logacheva, G. A. Bazykin, and A. S. Kondrashov. 2020. "Genomic Signatures of Recombination in a Natural Population of the Bdelloid Rotifer *Adineta vaga*." *Nature Communications* 11, no. 1: 6421. https://doi.org/10.1038/s41467-020-19614-y.

Valerio, T., B. Knop, R. M. Kreider, and W. He. 2021. "Childless Older Americans: 2018." Ed. U.S. Census Bureau. Washington, DC: U.S. Government Publishing Office.

van der Kooi, C. J., and T. Schwander. 2014. "On the Fate of Sexual Traits under Asexuality." *Biological Reviews Cambridge Philosophical Society* 89, no. 4: 805–819. https://doi.org/10.1111/brv.12078.

VanderLaan, D. P., R. Blanchard, H. Wood, and K. J. Zucker. 2014. "Birth Order and Sibling Sex Ratio of Children and Adolescents Referred to a Gender Identity Service." *PLoS One* 9, no. 3: e90257. https://doi.org/10.1371/journal.pone.0090257.

VanderLaan, D. P., Z. Ren, and P. L. Vasey. 2013. "Male Androphilia in the Ancestral Environment. An Ethnological Analysis." *Human Nature* 24, no. 4: 375–401. https://doi.org/10.1007/s12110-013-9182-z.

Vanderlaan, D. P., and P. L. Vasey. 2012. "Relationship Status and Elevated Avuncularity in Samoan Fa'afafine." *Personal Relationships* 19, no. 2: 326–339. https://doi.org/https://doi.org/10.1111/j.1475-6811.2011.01364.x.

Vasey, P. L., A. Foroud, N. Duckworth, and S. D. Kovacovsky. 2006. "Male-Female and Female-Female Mounting in Japanese Macaques: A Comparative Study of Posture and Movement." *Archives of Sexual Behavior* 35, no. 2: 117–129. https://doi.org/10.1007/s10508-005-9007-1.

Vasey, P. L., and D. P. VanderLaan. 2010. "Avuncular Tendencies and the Evolution of Male Androphilia in Samoan Fa'afafine." *Archives of Sexual Behavior* 39, no. 4: 821–830. https://doi.org/10.1007/s10508-008-9404-3.

Veale, D., S. Miles, S. Bramley, G. Muir, and J. Hodsoll. 2015. "Am I Normal? A Systematic Review and Construction of Nomograms for Flaccid and Erect Penis Length and Circumference in up to 15,521 Men." *BJU International* 115, no. 6: 978–986. https://doi.org/10.1111/bju.13010.

Veyrunes, F., P. D. Waters, P. Miethke, W. Rens, D. McMillan, A. E. Alsop, F. Grutzner, J. E. Deakin, C. M. Whittington, K. Schatzkamer, C. L. Kremitzki, T. Graves, M. A. Ferguson-Smith, W. Warren, and J. A. Marshall Graves. 2008. "Bird-Like Sex Chromosomes of Platypus Imply Recent Origin of Mammal Sex Chromosomes." *Genome Research* 18, no. 6: 965–973. https://doi.org/10.1101/gr.7101908.

von Besser, K., A. C. Frank, M. A. Johnson, and D. Preuss. 2006. "Arabidopsis HAP2 (GCS1) Is a Sperm-Specific Gene Required for Pollen Tube Guidance and Fertilization." *Development* 133, no. 23: 4761–4769. https://doi.org/10.1242/dev.02683.

Wagner, A. 2014. *The Arrival of the Fittest: How Nature Innovates.* New York: Penguin Random House.

Wallen, K., and E. A. Lloyd. 2008. "Clitoral Variability Compared with Penile Variability Supports Nonadaptation of Female Orgasm." *Evolution & Development* 10, no. 1: 1–2. https://doi.org/10.1111/j.1525-142X.2007.00207.x.

Wallis, M. C., P. D. Waters, M. L. Delbridge, P. J. Kirby, A. J. Pask, F. Grutzner, W. Rens, M. A. Ferguson-Smith, and J. A. Graves. 2007. "Sex Determination in Platypus and Echidna: Autosomal Location of SOX3 Confirms the Absence of SRY from Monotremes." *Chromosome Res* 15, no. 8: 949–959. https://doi.org/10.1007/s10577-007-1185-3.

Walter, M. R., R. Buick, and J. S. R. Dunlop. 1980. "Stromatolites 3,400–3,500 MYR Old from the North Pole Area, Western Australia." *Nature* 284 (April 3): 443–445.

Warnke, G. 2011. *Debating Sex and Gender.* Oxford: Oxford University Press.

Warrener, A. G., K. L. Lewton, H. Pontzer, and D. E. Lieberman. 2015. "A Wider Pelvis Does Not Increase Locomotor Cost in Humans, with Implications for the Evolution of Childbirth." *PLoS One* 10, no. 3: e0118903. https://doi.org/10.1371/journal.pone.0118903.

Washburn, S. L. 1960. "Tools and Human Evolution." *Scientific American* 203: 63–75.

Watkins, Graham G. 1997. "Inter-Sexual Signalling and the Functions of Female Coloration in the Tropidurid Lizard *Microlophus occipitalis.*" *Animal Behaviour* 53, no. 4: 843–852. https://doi.org/https://doi.org/10.1006/anbe.1996.0350.

Weaver, Lucas N., Henry Z. Fulghum, David M. Grossnickle, William H. Brightly, Zoe T. Kulik, Gregory P. Wilson Mantilla, and Megan R. Whitney. 2022. "Multituberculate Mammals Show Evidence of a Life History Strategy Similar to That of Placentals, Not Marsupials." *The American Naturalist* 200 (3): 383-400. https://doi.org/10.1086/720410.

Weigert, Anne, and Christoph Bleidorn. 2016. "Current Status of Annelid Phylogeny." *Organisms Diversity & Evolution* 16, no. 2: 345–362. https://doi.org/10.1007/s13127-016-0265-7.

Weismann, A. 1892. *Das Keimplasma: eine Theorie der Vererbung.* Trans. W. Newton Parker and H. Rönnfeldt. New York: Scribner.

Wellman, C. H. 2014. "The Nature and Evolutionary Relationships of the Earliest Land Plants." *New Phytologist* 202, no. 1: 1–3. https://doi.org/10.1111/nph.12670.

Welshons, W. J., and L. B. Russell. 1959. "The Y-Chromosome as the Bearer of Male Determining Factors in the Mouse." *Proceedings of the National Academy of Sciences of the United States of America* 45, no. 4: 560–566. https://doi.org/10.1073/pnas.45.4.560.

Werdelin, L., and A. Nilsonne. 1999. "The Evolution of the Scrotum and Testicular Descent in Mammals: A Phylogenetic View." *Journal of Theoretical Biology* 196, no. 1: 61–72. https://doi.org/10.1006/jtbi.1998.0821.

Werner, T., S. Koshikawa, T. M. Williams, and S. B. Carroll. 2010. "Generation of a Novel Wing Colour Pattern by the Wingless Morphogen." *Nature* 464, no. 7292: 1143–1148. https://doi.org/10.1038/nature08896.

White, M. F. 2011. "Homologous Recombination in the Archaea: The Means Justify the Ends." *Biochemical Society Transactions* 39, no. 1: 15–19. https://doi.org/10.1042/BST0390015.

Williams, George C. 1957. "Pleiotropy, Natural Selection, and the Evolution of Senescence." *Evolution* 11, no. 4: 398–411. https://doi.org/10.2307/2406060.

Wilson, D. S., and E. O. Wilson. 2007. "Rethinking the Theoretical Foundation of Sociobiology." *Quarterly Review of Biology* 82, no. 4: 327–348. https://doi.org/10.1086/522809.

Woese, C. R., and G. E. Fox. 1977. "Phylogenetic Structure of the Prokaryotic Domain: The Primary Kingdoms." *Proceedings of the National Academy of Sciences of the United States of America* 74, no. 11: 5088–5090. https://doi.org/10.1073/pnas.74.11.5088.

Wong, J. L., and M. A. Johnson. 2010. "Is HAP2-GCS1 an ancestral gamete fusogen?" *Trends Cell Biol* 20, no. 3: 134–141. https://doi.org/10.1016/j.tcb.2009.12.007.

Wood, W. B., ed. 1988. *The Nematode Caenorhabditis elegans.* Woodbury, NY: Cold Spring Harbor Laboratory.

Yadgary, L., R. Yair, and Z. Uni. 2011. "The Chick Embryo Yolk Sac Membrane Expresses Nutrient Transporter and Digestive Enzyme Genes." *Poultry Science* 90, no. 2: 410–416. https://doi.org/10.3382/ps.2010-01075.

Yin, D., and E. S. Haag. 2019. "Evolution of Sex Ratio through Gene Loss." *Proceedings of the National Academy of Sciences of the United States of America* 116, no. 26: 12919–12924. https://doi.org/10.1073/pnas.1903925116.

Yin, D., E. M. Schwarz, C. G. Thomas, R. L. Felde, I. F. Korf, A. D. Cutter, C. M. Schartner, E. J. Ralston, B. J. Meyer, and E. S. Haag. 2018. "Rapid Genome Shrinkage in a Self-Fertile Nematode Reveals Sperm Competition Proteins." *Science* 359, no. 6371: 55–61. https://doi.org/10.1126/science.aao0827.

Yoshizaki, N., W. Yamaguchi, S. Ito, and C. Katagiri. 2000. "On the Hatching Mechanism of Quail Embryos: Participation of Ectodermal Secretions in the Escape of Embryos from the Vitelline Membrane." *Zoological Science* 17, no. 6: 751–758, 8.

Yuan, C. X., Q. Ji, Q. J. Meng, A. R. Tabrum, and Z. X. Luo. 2013. "Earliest Evolution of Multituberculate Mammals Revealed by a New Jurassic Fossil." *Science* 341, no. 6147: 779–783. https://doi.org/10.1126/science.1237970.

Yuan, X., S. Xiao, and T. N. Taylor. 2005. "Lichen-Like Symbiosis 600 Million Years Ago." *Science* 308, no. 5724: 1017–1020. https://doi.org/10.1126/science.1111347.

Zaremba-Niedzwiedzka, K., E. F. Caceres, J. H. Saw, D. Backstrom, L. Juzokaite, E. Vancaester, K. W. Seitz, K. Anantharaman, P. Starnawski, K. U. Kjeldsen, M. B. Stott, T. Nunoura, J. F. Banfield, A. Schramm, B. J. Baker, A. Spang, and T. J. Ettema. 2017. "Asgard Archaea Illuminate the Origin of Eukaryotic Cellular Complexity." *Nature* 541, no. 7637: 353–358. https://doi.org/10.1038/nature21031.

Zhang, Qi, Michael Goodman, Noah Adams, Trevor Corneil, Leila Hashemi, Baudewijntje Kreukels, Joz Motmans, Rachel Snyder, and Eli Coleman. 2020. "Epidemiological Considerations in Transgender Health: A Systematic Review with Focus on Higher Quality Data." *International Journal of Transgender Health* 21, no. 2: 125–137. https://doi.org/10.1080/26895269.2020.1753136.

Zhou, Q., L. Clarke, R. Nie, K. Carnes, L. W. Lai, Y. H. Lien, A. Verkman, D. Lubahn, J. S. Fisher, B. S. Katzenellenbogen, and R. A. Hess. 2001. "Estrogen Action and Male Fertility: Roles of the Sodium/Hydrogen Exchanger-3 and Fluid Reabsorption in Reproductive Tract Function." *Proceedings of the National Academy of Sciences of the United States of America* 98, no. 24: 14132–14137. https://doi.org/10.1073/pnas.241245898.

Index

Sertoli cells, 31

sex determination: in *C. elegans*, 77–83; in *Drosophila melanogaster*, 68–72; environmental (ESD), 67; as a general phenomenon in animal development, xi, 65; genetic (GSD), 66–67; in mammals, 73–77, 127; in the platypus, 127; role of DM proteins in, 81–83

sex difference, 156, 166, 170, 172–173, 210

Sex-lethal (*Sxl*) gene, 71,81

sex role syndrome, 154

sex tendencies, 165–166, 169–171, 216

sheep, 85, 121, 133

Shine, Richard, 112

Shubin, Neil, x

Silurian, 42

Smithsonian Institution, 42, 243n9

soma, xi, 46–47, 51

SOX9 gene, 83, 86, 87, 89

sperm: competition, 102; as defining feature of maleness, vii–viii, xiii, 50–52, 83, 154, 165, 210; as derivative of germ cells, 28, 47; and the genome, 96, 128, 190; in hermaphrodites, viii, 53, 58–64, 67, 87, 89; in humans, 30–31, 34, 140, 151; in internal or external fertilization, 98–101, 105–106, 119, 132–133, 179; and role in sex determination, 67, 69, 79, 97; and temperature-sensitivity in mammals, 35

spina bifida, 168, 191

sponges, 56–57, 82

SRY gene (or SRY protein), 76–77, 80–85, 127, 170

Stanford University, 101

Stern, Curt, 74

Stevens, Nettie, 68–69

strangulation, 174

Strassmann, Beverly, 135

stromatolites, 5–6

Strongyloides papillosus, 85

Stryker, Susan, 198

Sturtevant, Alfred, 70, 75

Sullivan, Lou, 198

Swaab, Dick, 207

syncytins, 121

syngamy (or fertilization), 15–19, 24, 38, 46, 49, 51, 61, 66, 69, 87, 100–101, 106, 132, 136, 151, 168, 211, 234n1

Taliban, 162

testicular descent, 33–35

testis: of *Drosophila melanogaster*, 70; hermaphrodites, 61, 64, 82; of humans, 25, 30–31, 75–76, 80–81, 156; and sexual selection, 102

testis-determining factor (*Tdf*), 76

testosterone: in females, 170, 173, 235n14; in gender-affirming treatments, 37, 205; in human embryonic development, 31–32, 101, 169; in *K. marmoratus*, 88; in puberty, 36, 75

Thomas, Cristel, 96

Title IX, viii, 215

tomatoes, 87, 93–94, 138

Toxoplasma, 125

tra-1 gene (or TRA-1 protein), 79–82

TRA-2 protein, 79, 81

trans (gender identity), 13, 198–205

transcription factor, 76–77, 79–83

trans-exclusionary radical feminists (TERFs), 198

transformer (*tra*) gene (or Tra protein), 70–72, 81

transmission electron microscopy, 7, 78

Triassic Period, 110, 113

Trinity College (Hartford, CT), 112

TRPC2 gene, 72, 143, 146–147

Trump, Donald, 139

TRPV6 protein, 117

turtles, xi, 100–101, 104, 119, 126

twins, 188–189

United Nations Human Rights Commission, 124

United States Agency for International Development (USAID), 139

United States Census Bureau, 183

U.S. Supreme Court, 175, 198, 238n12

University of Amsterdam, 212
University of California, Berkeley, 71, 74
University of California, Davis, 27
University of Colorado, Denver, 125
University of Florida, 101
University of Illinois, 8, 112
University of Illinois, Chicago, 93
University of Lausanne, 118
University of Lethbridge, 186
University of Maryland, ix, 87, 117, 172, 210
University of Massachusetts, 198
University of Michigan, 113, 135
University of Minnesota, 88
University of Montreal, 158
University of Oregon, 91
University of Rochester, 7
University of Rhode Island, 126
University of Texas at Austin, 72, 222
University of Toronto, 186
University of Washington, 60
University of Wisconsin, 8
urethra, 32, 34, 37
uterus, 30, 33–34, 37, 115, 117, 119–125, 131–135, 151, 166

vagina, 33–34, 36, 102, 136–137, 174, 205
Vakhrusheva, Olga, 95
Vanderlaan, Doug, 186–187
vas deferens, 31
Vasey, Paul, 186–187, 222
Vikhanski, Luba, 170
viruses, 18, 121
vomeronasal organ (VNO), 142–146

Wagner, Günter, 120, 134–135
Warnke, Georgia, 213
Warrener, Anna, 125
Washburn, Sherwood, 125–126

Washington Post, vii, 204
water: conservation of in amniote eggs, 103, 114; emergence of terrestrial life from, 14, 38, 42, 51; formation in primordial earth, 4; as habitat for asexually reproducing organisms, 84, 86, 94; as limiting human resources, 194; and mammalian sense of smell, 144; as medium for fertilization, 58, 61–62, 98, 100; as medium for larval development, 99, 106; solubility of iron in, 5; solubility of oxygen in, 12; temperature of and mammalian physiology, 34–35, 109, 209
Weaver, Lucas, 113
Weismann, August, 47
Wesleyan University, 145
Western General Hospital (Scotland), 74
Woese, Carl, 8–9
Wolffian duct, 25, 29, 31, 37
Women's National Basketball Association (WNBA), 215

xol-1 gene (and XOL-1 protein), 79, 81, 239n29
XXY: females of *Drosophila melanogaster*, 69; male humans 75–76

Y-linked traits, 74, 77
Yale University, 102, 120, 135, 172
Yin, Da, 96
yolk, 48, 80, 105–106, 114–116, 118–120
Yoruba, 214

Zahavi, Amotz, 153
Zarkower, David, 80, 82
ZFY gene, 76
Zuckerkandl, Émile, xii
zygote, 15–17, 19, 21, 48–49, 51–52, 119

Milton Keynes UK
Ingram Content Group UK Ltd.
UKHW042009141124
451056UK00006B/54/J